Foreword

IT is both an honour and a great pleasure to have been invited to write a foreword to this excellent book which continues the work of Professor Piaget and myself and places it in an educational context. We have been following with great interest the numerous experiments reported by Dr Lovell, and it is now very satisfactory to have this full account which is at once complete, clear and attractive.

I should like first to congratulate Dr Lovell on having been able to do something we have not succeeded in doing in French: on having written for school teachers a psychological work that is so accurate and so scientific. Most French works of this kind go no further than a rather superficial popularization. Dr Lovell has, on the contrary, a very different idea of the function of school teachers. He expects of them not only a knowledge of the subjects to be taught but also an understanding both of their psychological structure and of the way in which scientific ideas, beginning as spontaneous operations, are elaborated in the pupils' minds. This is work in a tradition which we have not been able to establish on the Continent and which makes us just a little envious, a genuine collaboration between psychologist and teacher. It is only fair to recall in this connection the excellent work of Professor Peel, Miss Churchill and Dr Dienes, etc.

Dr Lovell does not accept any of our findings or take over anything from our work without the most scrupulous checking, and he deserves very warm congratulations. I must admit at the same time that we are always just a little worried when we wait for the results of such verifications, because they deal with researches that are carried out not only in a rather different school setting but also, and above all, with methods that differ from ours. We have used a very flexible method of exploration, because at the stage when we explore childrens' ideas as yet unknown to us, it is essential to proceed in the most adaptable way. On the other hand, at the level of experimental checking at which Dr Lovell is working, a standardized method is necessary and we know from our experience how delicate a matter is the translation of the first method into the second, and the

5

risk there is of losing a great number of nuances and even of falsifying certain aspects of the thought of the child. It is all the more reassuring to see that Dr Lovell's results and our own agree perfectly in broad outline. There are the same beginnings, the same formative mechanisms, the same stages and nearly the same ages.

Even where Dr Lovell suggests some differences between his findings and ours we see convergence. For example, Dr Lovell has rediscovered a point on which Piaget has always insisted, that a concrete operation is not generalizable into all contexts but remains specific to a particular context. Thus we, too, found that the child formed his first ideas of conservation of quantity in relation to a liquid poured from one vessel to another, and only then in relation to the deformation of the ball of plasticine. Similarly Dr Lovell rightly insists on the fact that reversibility of thought is a necessary but not sufficient condition—that 'operativity', when completed, is one with the structure of the group. Thus Smedslund has found at the level of the first beginnings of their construction a very significant correlation between reversibility and transitivity in the formation of ideas of conservation of weight. It is true that this concerns the transitivity of equals and is not necessarily the same if one compares conservation with the transitivities of inequalities, as Dr Lovell does. Finally, it goes without saying that the topological structures to which we alluded are not the general structures of the topologists in pure geometry. We are only concerned, in children, with certain relations which correspond to a topology of Euclidean space. But there remains the interesting fact that the child begins by abstracting, by means of tactile and visual exploration, certain qualitative and topological aspects of space only to discover later metrical relations, that is to say, relations which correspond to Euclidean geometry in the historical sense.

In conclusion I congratulate Dr Lovell on having made a simple analysis of the structures of thought, which makes easier the reading and understanding of work in this field.

It only remains for me to wish this excellent work the success it deserves. There can be no doubt that it will be warmly welcomed both in English-speaking countries and on the Continent.

BÄRBEL INHELDER

Preface

THIS book has been written in the hope that it will help all those interested in children to think more clearly about certain mathematical and scientific concepts. There are a number of excellent textbooks that deal with more advanced mathematical notions from an elementary standpoint, but to my knowledge there is no elementary text dealing with current thinking about some of the basic ideas, and the possible stages through which children pass in arriving at them.

Although this is not a book about teaching method as such, it may encourage students and teachers to reflect upon the activities in which they engage in their teaching, and to see the possible value and likely limitations of these activities in a fresh light. Often we do not appreciate why we do certain things in the classroom. Readers may not agree about the choice of concepts discussed, but authors must be selective if books are to be kept to a reasonable size and price.

I have drawn heavily upon the works of Professors Piaget and Inhelder and their students, and this opportunity is taken of paying a warm tribute to them for the ingenious experiments that they have provided, and for the light they have thrown upon the stages of child development. Although there are a number of points on which I find myself in disagreement with the Geneva school, I strongly urge readers to study the books written by Piaget and Inhelder and to repeat for themselves some of the experiments described. Without exception students working under my direction, who have now carried out about ten thousand experiments of the type first suggested by Piaget and/or Inhelder, have acknowledged the insight that the Geneva school has provided into the stages of child thinking. My experience leads me to believe that all teachers in training should have the opportunity of observing children undertaking some of the Piaget-type experiments.

In the last chapter but one there is a short and simple treatment of the number system. It is included for two reasons. First, the extension of the rules of arithmetic from natural to other numbers is a good example of the mathematical process of generalization. Second, it gives in simple language a short outline of the system; and a recent publication of the Association of Training Colleges and Departments of Education (*The Mathematical Ideas and Principles which underlie the construction of Main Courses in Mathematics*: 1958), recommends that an understanding of *number* can be considered a necessary part of the equipment of all engaged in teaching mathematics from infant level upwards. Some readers will consider this chapter a difficult one; but it will amply repay studying.

The National Foundation for Educational Research is now carrying out a long-term comparative study of the mathematical competence of primary school children who have been taught with the Stern or Cuisenaire materials on the one hand, and others who have followed various 'active' approaches or who have been subjected to the so-called 'traditional' methods, on the other hand. It is hoped that this book will, among other things, help to explain the rationale of the methods whose relative effectiveness is being studied in the N F E R investigation.

It must be realized by teachers and others responsible for the upbringing of children that new and sometimes complex ideas are now evolving as psychology, and even physiology, neuro-surgery, and genetics give us new information regarding the factors that control the development of the human being. If they are to remain well-informed teachers and others concerned with the growth of children must come to grips with these ideas.

I wish to thank all those with whom I have discussed the topics covered in this book for the stimulation that I have received from them. These discussions have taken place with teachers who have had greatly differing opportunities for using teaching apparatus and materials in mathematics—ranging from teachers in Britain who could avail themselves of any apparatus, whatever, that could be bought or made, to teachers in the more inland parts of Brazil all of whose teaching apparatus had to be made from the bamboo of the forest.

K L

Contents

CHAPTER ONE

Concept Formation

THE PERCEPT

SINCE we are going to deal with important mathematical and scientific concepts we must begin by discussing the term *concept* itself. We must try to understand what is meant by this word, and must examine, very briefly, the possible ways in which a child may arrive at a concept. But before we can do this we must discuss the meaning of the word *perception*.

When stimuli from the external world in the form of sight, sound, touch, and smell, enter the central nervous system via the appropriate sense organ, they are subjected to a process of filtering. That is to say, it seems as if the system is of limited capacity, so that some kind of selection of signals must take place, especially if the signals are complex. The factors that determine the selection seem to be the nature of the incoming stimuli, the likelihood of their occurrence, and conditions within the individual, such as expectancy and his needs. After this selection has taken place, the signals reach the cerebral cortex and the related areas in the mid brain, and we experience certain sensations.

The *interpretation* that is given to these incoming signals, —that is, our *perception* of the external world—depends on more than the sensations that reach the cerebral cortex and mid brain. Perception results from the reinforcement of these sensations by past experience, ideas, imagery, expectancy and attitude. We do not, of course, recognize the receipt of sensations and the activity of perception as separate processes. Note, however, that learning plays a great part in the interpretation we give to these sensations. Perception is therefore likely to be affected by our ways of thinking, by our attitudes, emotional states, expectation or desire at a given moment, so much so that we sometimes perceive, quite falsely, that which we have been expecting to perceive. Again, it must be remembered that perception, as distinct from imagery, results from immediate contact with the relevant part of the environment.

THE CONCEPT

It is now possible to discuss ways in which concept-formation may take place. We cannot be definite, because we know, as yet, very little about the way in which children arrive at their concepts. Indeed, it seems that different children arrive at the same concept in different ways. When we hear the word 'bird' spoken, or see the word in print, we do not think of all existing kinds of bird, ranging from the barnyard fowl to the swallow. For the normal person, the word implies a *class* * of animals having feathers and two legs, most members of which can fly. He has picked out some common properties in certain animals and to those that have these properties he has given the name 'bird'. To generalize, concepts enable words to stand for whole classes of objects, qualities, or events, and are of enormous help to us in our thinking.

We shall begin by discussing the rather traditional view of concept-formation. When the child forms a concept he has to be able to *discriminate* or *differentiate* between the properties of the objects or events before him, and to *generalize* his findings in respect of any common feature he may find. Discrimination demands that the child should recognize and appreciate common relationships, and differentiate between these and other unlike properties. For example, the common feature among a number of circles of different diameters, made of different materials and of different colours, is the roundness of the circle, and the recognition of this feature in all the objects constitutes a major step forward in concept-formation in this case. On the other hand, the variable characteristics, such as diameter, colour, and material, are ignored; for they contribute nothing in this instance. Some people, however, prefer the term *abstraction* (in the sense of 'taking from') rather than *discrimination*. In addition to abstraction or discrimination, *generalization* occurs, by means of which some notion of the

* A class may be formally defined as all those entities having a certain property; e g man.

concept has been developed. The partly-defined concept stands now as a hypothesis (e g 'It's a triangle'; 'It's an invertebrate') which is tested by trying it on fresh specimens.

The child begins with percepts. But from infancy he begins to discriminate, abstract, and generalize about environmental data. He does not, of course, understand or control this process of abstraction or 'taking from', nor is he aware of it at first; as far as he is concerned, it simply happens. As he grows older, there is greater awareness and deliberation. If he encounters a variety of stimulating experiences, abstractions and generalizations are likely to proceed more readily, provided the experiences are matched with the child's neurophysiological development. The sequence is: perception—abstraction—generalization.

Abstraction and generalization are essentially mental processes; they are carried out in the mind. Adults can arrange an environment which may help them forward but the child himself makes the jump from the percept to the concept.

A concept may be defined as a generalization about data which are related; it enables one to respond to, or think about, specific stimuli or percepts in a particular way. Hence a concept is exercised as an act of judgment. Concepts seem to arise out of perceptions, out of actual acquaintance with objects and situations, and through undergoing experiences and engaging in actions of various kinds.

Concept formation is also likely to be aided by memories and images. For example, as the child is building up his concept of *transport* he is helped by memories of car and train journeys, and by images of ships seen at the seaside and of aeroplanes flying over his home town. As the concept becomes more completely generalized he can speak of *transport* instead of cars, trains, ships and aeroplanes.

It is important to stress that in children concepts do not usually develop suddenly into their final form. Indeed, concepts widen and deepen throughout life, as long as the brain and mind remain active, and providing prejudice does not narrow categorization.

Sometimes in concept-formation there is a certain amount of trial and error in order to determine if a new specimen fits into an existing hypothesis. It is certain, too, that reasoning is

13

often involved when concepts are being formed, because selection of the relevant and rejection of the irrelevant must take place. Vinacke (1952), an important worker in this field, claims that in adults both abstraction and generalization are more dependent upon motivation, and are more conscious and controlled than they are in the child. This is to be expected. Further, at adult level, the concepts that we have acquired have an effect on our perceptions; there is, as it were, a feed-back effect.

Language and mathematical symbols certainly play a part in concept formation for they enable the individual to pin down and clarify concepts, or act as a frame of reference. Moreover, concepts enable us to communicate our thoughts to others in written and spoken language which is thus of particular help in aiding the child to develop and discuss concepts like *honesty* and *automation*. At the same time it is Piaget's view that while language aids the formation and stabilization of a system of communication constituted by concepts, it is in itself insufficient to bring about the mental operations that make systematic thought possible. On this view, language translates what is already understood; for language is essentially a symbolic 'vehicle' for thought. It is certain that not only very young children but even rats can develop some notion of, say, *triangularity* because they will respond to quite dissimilar triangles in a similar manner. So this, and other simple ideas, can be partly developed without verbalization.

Both children and adults may have developed a concept good enough for working purposes in everyday life—and in the case of the adult good enough for his professional or technical life—and yet be unable to define the concept in verbal terms. This is very frequent, and is not necessarily due to lack of vocabulary. On the other hand, teachers are often deceived because children can use the appropriate word and yet they have no idea of the related concept.

At the nursery and infant school stages a child's concepts are still fragmentary and limited. He frequently does not see an object as one instance of a class or category. There has been, as yet, insufficient abstraction and generalization; the concept is not fully developed; so that he is likely to think of a thing in terms of a concrete situation, that is, he defines it descriptively. That is why we hear young children saying, 'I use a knife and fork to eat with', or 'A nose is to blow'. When the concept is better developed, he need no longer report concrete uses, odd examples, or isolated experiences.

So far it has been implied that in his intellectual development the child moves essentially from the concrete to the abstract. But in many instances it may not be so. Some people think that in the young child there is little discrimination or differentiation. It is true that the infant cannot fail to make certain discriminations, as between a bright light and darkness, or between a loud noise and a whisper. Such discriminations as cannot be avoided divide the perceived world into a small number of very large categories. As intellectual development proceeds, discrimination increases, and as the number of distinctions gets larger, the number of categories increases, and the smaller and more concrete they become. Brown (1958) argues strongly along those lines, and points out that the fact that the young child will use the word *daddy* to denote all men, does not show that he is deficient in abstracting ability, but that he has to employ large categories at first.

A child's use of language oscillates between these viewpoints. The words the adults first teach children are those which categorize the object in the way most useful for everyday life. For instance, the child uses the word *fish* first, and words like *salmon* and *plaice* later. Likewise, he might use *money* before *penny* and *shilling*. On the other hand, he certainly uses the words *milk* and *water* before the word *liquid*, and *knife* and *spoon* before *cutlery*. Brown argues that he first acquires the concepts which adults think are of maximum value to him. His verbal concepts may then move to a greater degree of abstraction or a greater degree of concreteness, so that the sequence in which a particular set of terms is used is not due so much to his intellectual preferences as to what adults consider to be the most useful naming practice. To some extent adults impose their own cognitive structures on children.*

* For a relevant paper on classification see Annett, M, 'The Classification of Instances of Four Common Class Concepts by Children and Adults', *Brit J Educ Psychol*, 1959, **29**, 223–236.

Piaget and Inhelder (1959) show the growth, from 4 to 10 years of age, of the child's ability to classify objects. Their findings have been broadly confirmed by the writer (Lovell *et al* 1962). This ability to classify seems to depend on the capacity to compare two judgments simultaneously, and it may well arise out of the child's increasing ability, from the age of a few weeks, to coordinate retroactions and anticipatory processes. Simple forms of classification can be carried out independent of language, but the latter is necessary for more complex forms of classification since language clarifies the category and focuses attention on it. We have also confirmed Piaget's view that it is as easy for the child to classify objects using tactile and kinaesthetic perception (objects felt but not seen) as by visual perception.

Price-Williams (1962) has provided evidence regarding the growth of the ability to classify in a primitive society. Studying bush (illiterate) and primary school children in the Tiv tribe in Nigeria he compared their ability to classify and sort materials familiar to them. He found no differences between the two sets of children over the approximate age range of $6\frac{1}{2}$–11 years and readers should consult the article itself to note the similarites and slight differences between these children and European children in the ability to classify. It seems, indeed, that the ability to discriminate and classify is something fundamental to the growing organism.

A good deal has been said about discrimination and generalization but it has also been admitted that we do not really know how these occur. Bartlett (1958) has put forward some suggestions about the process of generalization in adults. To what extent these suggestions hold in respect of similar functions in children we do not know, although there is no reason why they should not apply. In Bartlett's view, when generalization takes place in any kind of formal or experimental thinking, the mind must make an *active* search for all the points of similarity between the ideas in the data before it. That is discrimination. The search carries on until the mind is satisfied that the points of agreement that are consistent with the observed differences, have been recognized in such a way that the *number* and *order of steps* (i e the direction) in the thought-sequence is the same for all the instances. The agreements noted are thus treated as belonging to a system, and they can be recognized in any other example.

Bartlett points out that this view runs contrary to a widely-held view about generalization, namely, that from the data within the individual's experience, likenesses and differences automatically separate themselves out. The likenesses are then given names and are termed *qualities* or *properties*. Again, generalization, and transfer of the results of practice and training, are usually thought of as two aspects of the same kind of mental process. But if the former necessitates active search, transfer of training cannot be guaranteed merely because there are points of likeness between data. It seems that for transfer of training to occur there must be active exploration that definitely *seeks* to make use of the features of the situation. We cannot be sure that transfer of training will take place merely because appropriate instances are brought together.

We must also note the difference that, in Bartlett's view, exists between generalization in formal and experimental situations on the one hand, and in everyday life on the other. In the former, instances are carefully explored; in the latter, one or two instances are 'savoured' but not explored.

It has already been said that a concept consists of a generalization about data which are related. The wider and deeper the concept, the greater the degree of generalization that can be made about the relationships. Unfortunately, little is known about the exact ways in which concept-formation can be speeded up although we know that environmental conditions are of importance. Lovell (1955) showed that adolescents and young adults from a stimulating background were, in ability to categorize and form fresh concepts, superior to those from less favourable backgrounds—this after making allowance for general intelligence or academic aptitude. Churchill (1958) showed that infant children who had had opportunities to play with certain materials might develop certain mathematical concepts quicker than a control group who were not given such opportunities.

THOUGHT ARISING FROM ACTIONS

So far no explicit mention has been made of the view that all *thought* is dependent upon actions. By *thought* we mean *a connected flow of ideas directed towards some definite end or purpose.* And whenever in this book the words 'thought' 'think' or 'thinking' occur they are used in this sense. Piaget, as we shall see later, is certainly of the view that thought arises out of actions, and mathematical concepts arise out of the actions the child performs with objects and not from the objects themselves. During the first two years of life the child slowly builds up a repertoire of actions, and accumulates much experience of their effects, while at the same time his central nervous system is maturing. According to Piaget, an ordinary child can, by about his second birthday, work out how he is going to do something before he does it, provided the situation is simple and is familiar to him. He can represent to himself, the results of his own actions before they occur. This is the beginning of true thought, since actions have become 'internalized'. This is a skill that, as far as we can judge, the cleverest animal rarely attains.

The fundamental skill that underlies all mathematical and logical (internally consistent) thinking is the capacity for 'reversibility'; i e the permanent possibility of returning in thought to one's starting point. In Piaget's view, reversibility has its origins in actions starting in the first few weeks of life when the child pushes, say, a toy away from him and then pulls it back. Once again there develops the ability to co-ordinate retroactions and anticipatory processes. Presumably it is genetic differences and experience that determine individual differences in the organism's capacity to 'search' the environment and co-ordinate *schemata*. Later it will be shown that there is a great increase in the power of reversibility of thinking from 7 years of age onwards. Children cannot learn from mere observations. Their own actions have first to build up systems of mental operations, and when these become co-ordinated with one another children can begin to interpret the physical world. Thus Piaget's view is that there is no direct dependence of conceptual development on perceptual development. Conceptual development is essentially the development of the schemata of action in which perception plays a part.

Piaget is by no means the only person who has maintained that action is the basis of thought. Sherrington, the distinguished physiologist, was of the opinion that mind seems to have arisen in connection with the motor act. Meredith (1956) suggests that primitive man learnt to operate manually before he got far with any mental operations, i e actions performed in the mind. He thinks that through man's proprioceptive * sense the manual operation associated with his primitive crafts became 'internalized'. Early man found himself with visual and other images, but through his actions he acquired a stock of 'operational' or 'manoeuvrable' images—i e thoughts. The last-named images were dynamic in character, the others were not, and so early man found himself not only performing actions but envisaging actions and their results. In this way he began to think.

Hence it seems that as a result of the interaction of the organism with the physical environment, the former constructs certain concepts, e g number, time etc, and developes certain forms of internally consistent thinking that maximises his chances of understanding or 'making meaningful' his environment and predicting within it. The human organism thereby brings about maximum adaptation in relation to the matter energy system. This adaptation or 'equilibration' is highly influenced by practice and experience which brings out latent contradictions in a situation, and thereby initiates a process of inner organisation within the child.

For Piaget the type of concept that develops depends essentially on the level of abstraction or dissociation of which the child is capable, and this, in turn, depends upon the quality of the sequences of action in the mind, termed *schemata* or *schemas* that the child can elaborate. From about two years of age the child begins to form what Piaget calls a *pre-concept*; that is to say, the child dissociates objects from their properties on the basis of their behaviour, e g bread knife from penknife. But from about 7 years of age the child increasingly develops new and more complicated schemas. He can now 'look down on his schemas' or 'turn round on his schemas'. He becomes *aware* of the sequences of action in his mind, and he can see the part played by himself in ordering his experience. Thus it becomes possible for the child to build the *concepts* of class, relation, number, weight, time, etc. But it is only concepts that can be derived from contact from first hand reality that can be elaborated by him.

From 12 years of age in able pupils, and from 13 to 14 years of age in average ones, more advanced types of concept can be built for the schemas now available are more complex than those available in the junior school period. It is now possible for the pupil to structure and coordinate actions (in the mind) upon relations, which themselves result from the coordinations of actions. An example will make this clear. In the concept of *proportion* the adolescent has to build a relationship between, say, 2 and 7; also between, say, 6 and 21; and then establish an identity relationship between the two previously established relationships. The logical structure of such a system is exactly parallel to that of a statement of proportionality.

CONCEPT FORMATION AND MATHEMATICS

So far we have been dealing with concepts in general. It was necessary to have this discussion to put the problems of concept-formation in proper perspective. Mathematical concepts are one class of concepts; they are generalizations about the relationships between certain kinds of data. When dealing with, say, the natural numbers (1, 2, 3, 4 . . . etc) the child has to move from percepts (arising from the environment) and actions to the concept, and the methods that the teacher uses aid this process to a greater or lesser extent. If the concept of the natural numbers does not ultimately develop in the child, if they do not exist in his mind divorced from particular things, apparatus, actions, or circumstances, the mental operations or manoeuvres that he can perform with them will be limited accordingly.

In the report of the Harvard Committee in 1945 entitled *General Education in a Free Society* (quoted by Newsom, C V, in *Insight into Modern Mathematics*; Washington, 1957) occur these words, 'Mathematics studies order abstracted from the particular objects and phenomena which exhibit it, and in a generalized form'. Here we find a truth which it is important for all teachers to grasp, namely, that *mathematics is a mental*

* The proprioceptors are sense organs that continuously register changes in length, tension, compression etc of the body. Through these we become aware of, and judge, the position of our arms, legs, and other parts of the body, both in relation to one another and the outside world.

activity—and writing on paper is merely an aid. Furthermore, there are whole systems of concepts involved in mathematics, e g numerical and spatial ones, and mathematics studies the relationships between these and the mental operations (or manoeuvres) that can take place between them. To help the child to develop his mathematical concepts we must also teach the language and symbols of mathematics. However, the grasping of mathematical concepts is not the beginning and end of mathematical ability. Such ability demands besides the understanding of concepts, knowledge of mathematical language, symbols, methods, and proofs. Some of these have to be learned, retained and reproduced; they have to be combined with other concepts, symbols, methods, and proofs, and jointly 'manoeuvred'; and they have to be used in solving mathematical tasks. The individual will not get very far in his mathematical thinking unless he has the concepts, although *he may not be able to formulate a clear definition of the concept in verbal terms.*

It is often argued that if children were made acquainted with mathematical ideas earlier than they are, and consequently with the language and symbols, concepts would develop earlier. There is some truth in this. It is true as Margaret Mead * has pointed out, that in some primitive societies people can only count up to 20 to 25. Even the brightest of these may not count beyond 50, whereas in our society people with relatively low intelligence quotients can develop number concepts and use some algebraic concepts, because our society is so much more impregnated with mathematical ideas. One must never forget the unconscious influence of 'the spirit of the times', that hidden persuader of insight. So the further mathematical ideas obtrude into everyday thinking the more will children imbibe them. On the other hand, there is undoubtably an age limit in children, below which, owing to immaturity, they cannot develop any notions of a particular concept. This limit differs from child to child.

Teachers differ about the kind of approach most likely to be of help in developing mathematical concepts, especially in

* Mead, M. In *Discussions in Child Development*. Vol 3. London: Tavistock Publications, 1958, pp 61–62.

the infant and junior stages. Some think when dealing with the concept of natural numbers, it is best to make use of the materials of everyday experience (e g The Mathematical Association's report on *The Teaching of Mathematics in Primary Schools*, Chapter Two) rather than to utilize special pieces of apparatus in order to make clear specific points and stages. On this view the child abstracts and internalizes or intellectualizes the problem, or becomes aware of the significance of his own actions—however we term it—through a very wide range of activities and experiences, so that there is little need for direct teaching. Others maintain that a more specifically guided approach, using such apparatus, say, as that of Cuisenaire, Montessori or Stern, is necessary, to augment his other experiences. When this procedure is followed the child has to handle material, answer questions, and make choices in order to formulate, in consciousness, the relationships and properties of the material before him. Ultimately, the relevant concept has to exist as an abstract concept if it is to be fully operational or manoeuvrable in the mind. For example, at a later stage, the child may derive the idea of the negative integer 'minus two' from discussion about temperatures above and below freezing points, heights above and below sea level, or points to left or right of an origin, but unless 'minus twoness' can exist in his mind as an abstract concept, directed numbers will never be fully operational. He cannot derive this concept from natural materials, as he can derive his concept of 'twoness' from objects in the environment, because negative integers correspond to nothing in the external world, although they play a great part in mathematics. One is not, of course, belittling the use of illustration indicated above, but stressing that the concept of negative integers must exist without being tied to particular instances or environmental situations.

In Western Europe and the U.S.A. many girls at the secondary school stage fall behind boys in their understanding of mathematics. Saad's study (1957) yielded some extremely interesting data regarding the discrepancy between the two sexes' understanding of concepts and principles in fourth year grammar school pupils. The schools surveyed were mixed, and so both sexes were taught by the same teachers. Saad gave some numerical data throwing light on problems which had previously been discussed only in more general terms. He found that by the fourth year, girls had fallen behind boys in understanding concepts and principles in all branches of mathematics and that the greatest difference was in geometry. Does this indicate a difference in innate potential between the sexes, or does it result from the differing upbringing and training accorded the sexes which, almost from birth, are treated differently in our society? This question cannot be answered yet. It is true that boys seem to play with bricks, blocks, etc (making their own free choice) more frequently than girls, by 2–3 years of age. Moreover, there appears to be a difference, in favour of the boys, in operations involving the imaginative manipulations of shapes in the mind. It is also known that opportunities for manipulating objects and for tactile exploration aid the individual in representing to himself these spatial relationships. If our society encourages such activities more in the boy than in the girl, it could, presumably, cause discrepancies in understanding later on. But the solution to this problem is not yet known.

THE VIEWS OF DIENES ON CONCEPT FORMATION

Dienes (1959), in a highly original study of mathematical concept-formation in children, makes some suggestions about the distinctions between *analytic* and *constructive* thinking (see also Bruner, J S, 1960, p 57/58). In the former, the individual uses logical thought so far as he is able, so that his concepts are clearly formulated and defined before he uses them. In the latter, he first gets an intuitive perception (i e perception not based on reasoning) of something which is not fully understood. This rather vague perception urges him on to constructive or creative effort to confirm the intuition by logical argument. In Dienes's view constructive thinking develops before analytic thinking, although both are required in mathematical and scientific studies. The ideas of Dienes are clearly relevant to the discussion in Chapter Two, and his teaching materials are briefly described in Chapter Four. He has also produced some evidence that there is a sex difference between the children's ability to do the tasks he set them, and certain of their

23

personality characteristics. The study cannot tell us whether sex differences in mathematical concept-formation are due to heredity or to culture pattern. It is more than likely, however, that personality differences cut across the subdivision of the population into two sexes, and the first problem to be attacked might be to determine what types of concept organization occur in different types of personality.

REFERENCES

BARTLETT, F C (1958). *Thinking*. London: Allen and Unwin.

BROWN, R (1958). 'How Shall a Thing be Called?' *Psychol Rev*, 65, 14–21.

BRUNER, J S (1960). *The Process of Education*. Harvard University Press.

CHURCHILL, E M (1958). 'The Number Concepts of the Young Child'. *Researches and Studies*, University of Leeds Institute of Education, 17, 34–49; 18, 28–46.

DIENES, Z P (1959). *Concept Formation and Personality*. Leicester: Leicester University Press.

LOVELL, K (1955). 'A Study of the Problem of Intellectual Deterioration in Adolescents and Young Adults'. *Brit J Psychol*, 46, 199–210.

LOVELL, K, MITCHELL, B and EVERETT, I R (1962). 'An Experimental Study of the Growth of Some Logical Structures'. *Brit J Psychol*, 53, 175–188.

MEREDITH, G P (1956). 'Mathematics and Mind'. *Mathematical Gazette*, 40, 103–108.

PIAGET, J and INHELDER, B (1959). *La genèse des structures logiques élémentaires*. Neuchatel: Delachaux and Niestlé. English edition translated by E A Lunzer, London: Routledge and Kegan Paul, 1964.

PRICE-WILLIAMS, D R (1962). 'Abstract and Concrete Modes of Classification in a Primitive Society'. *Brit J Educ Psychol*, 32, 50–61.

SAAD, L G (1957). *Understanding in Mathematics*. University of Birmingham Institute of Education.

VINACKE, W E (1952). *The Psychology of Thinking*. London: McGraw Hill.

The Problem of the Logical Basis of the Natural Numbers and of Mathematics Generally

THE numbers 1, 2, 3, etc which we use so often in everyday life are called the *natural* numbers, because it is generally felt that they have, in a philosophic sense, a natural existence quite independent of man. Later we shall see that there are other numbers, such as negative integers (-1, -2, etc) and fractions, which, with yet further kinds of numbers, make up the entire number system. These other numbers, however, are regarded by some as the constructions of men's minds, since negative integers, for example, do not correspond to anything in reality. Kronecker said 'God created the natural numbers; all other numbers are man's handiwork'.

The natural numbers are an abstract concept, and as Bertrand Russell has pointed out, it took man a long time to recognize the quality of 'twoness' both in a brace of pheasants and in two days. Man's mind has in some way, after examining the phenomena of his environment, derived from it the concept of the natural numbers. While it is true that the concept of the natural numbers is easier than that of other numbers, it is nevertheless an abstract concept. From this it follows that natural numbers are entirely independent of the words or symbols we use to represent them. Although a study of the ways in which man has represented numbers in the past is a very interesting one, it is sufficient for our present purpose to note that the Hindu-Arabic system of representation is now

almost universally used to represent numbers. The excellence of this notational system lies in the fact that the ten symbols can represent all numbers, the different positions of the symbol indicating the different values that it can have.

It is not generally realized, however, that although the whole edifice of modern mathematics is built upon the concept of the natural number, this concept remains something of a mystery. To make clear the nature of this problem we will first discuss the way in which it seems likely that primitive man arrived at some understanding of natural numbers. Wheat (1937) has sketched the possible stages through which primitive man passed; in the next three paragraphs use will be made of his views.

When early man returned to his abode he may have wanted to tell his family of his experiences and to describe the animals he encountered. He may well have used terms corresponding to our 'many' and 'few', a large group for example, being described as 'many, many'. Out of these and other experiences there arose a need for quantitative exactness. The use of names for animals or objects helped at this stage, providing a man's possessions were few. If he had, say, three sheep, he would have names for each, and he could tell if they were all present. On the other hand, if he had no names for the sheep in his flock he had only a vague idea whether or not they were all present. Later, he hit upon the device of matching the objects of one group with those of another group. For example, if flint axes were handed round to a group of men, it would soon be clear if there were enough axes or too many or too few. This one-to-one correspondence was important for the later concept of number, and it also led, no doubt, to terms like 'more', 'less' 'as many as'. This matching was doubtless accidental at first, but later it was deliberate in the sense that the group of objects was matched against a model group, e g wings of a bird, paws of the lion, fingers of the hand. Then when he spoke about his group, he said that he had seen as many bears as there were toes on his feet.

But to deal with large groups he eventually resorted to tallying. A notch was cut in a stick for each object, or a pebble put aside for each animal, so that the herdsman could check his sheep against a number of stones. The stones and sheep were quite dissimilar, but each sheep and each stone represented a unit and there was a one-to-one correspondence between them. Tallying was very useful and a marked step forward, but even when primitive man had 'tallied', he still could not think of, or name, the number.

The uses of model groups and tallying were limited. Early man's first concept of fiveness would be in terms of the number of fingers on the hand, not the abstract 'five'. It was a big step, intellectually, to move from words that stand for model groups to words that stand for abstract numbers. Exactly how this took place we do not know. The associated sensory impressions—the perceptions—of a one-to-one correspondence, and/or the actions involved in setting up this correspondence, were most helpful in helping man to arrive at the concept of the natural numbers; but they may not have been in themselves sufficient. These perceptions and actions, perhaps, did not in themselves bring in the idea of numbers, but they increased greatly the chance of the concept being formed. There had to be in early man, as in every child today, an intellectual jump, to the idea of an abstract 'twoness' and 'threeness'.

How, when objects are matched in front of the child, or primitive adult, is insight gained into the idea of the natural numbers? How does the child recognize the quantity of 'three-ness' in two or more groups of 3 (three apples, three marbles etc)? No definite answer can be given to these questions, as mathematicians are not agreed among themselves. Some, for example the French mathematician and philosopher H Poincaré, believe that at this point the idea of a series of natural numbers becomes clear to everyone. Such people believe that the concept of the natural numbers is the result of a primitive intuition at this stage.

From making up and seeing groups of 2, 3 or 4, the child recognizes the inherent twoness, threeness or fourness in number, and actions and perceptions are the preliminary steps in the acquisition of the concept of cardinal numbers. While one, two, three and four have their origin in action and group impression, and are the names for these groups, the higher cardinal numbers are a substitute in a conceptual way, for

27

groups that cannot be known by simultaneous apprehension of their constituent objects. For example, the child (or adult) can have an intuitive or perceptual knowledge of 3, but only a symbolic knowledge (to use a term employed in logic) of, say, 87,925. The larger cardinal numbers are more abstract than the first four numbers.

Exactly how the mind acquires this intuitive knowledge of the first three or four cardinal numbers is not known. We do not yet understand the mystery of the human mind's strange faculty for perceiving analogies and formulating categories, and we cover our ignorance by coining words like 'insight' and 'intuition'. And our ignorance in this respect is likely to remain until there are advances in neurophysiology which enable us to have a better understanding of the physiological basis of the classifying mechanisms.

But there are other people who do not agree with Poincaré and who think that an understanding of numbers is based wholly on logic. This would imply that certain logical concepts have to be acquired by the child before a grasp of numbers is possible. Among such people are Bertrand Russell, A N Whitehead, and J Piaget, although the views of the last named do not agree with those of the other two. The views of Russell and Whitehead will be given very briefly in the next paragraph and those of Piaget given in more detail later.

Figure 1

In Figure 1 we see a set of three triangles and a set of three squares, such sets being called *classes* in logic. A *class* is defined as any collection of entities, which may be concrete entities e g elephants, chairs, men; or symbols (as in this case); or abstract entities such as classes themselves (e g the class of all good men, the class of all honest women). The entities (here symbols of triangles and rectangles) are called the members of the respective classes. Again, in this example there exists a one-to-one correspondence between the triangles and the rectangles, one triangle corresponding to one rectangle. Accordingly the two classes are said to be similar, and the common property possessed by all classes similar to either or both of the two classes is what is generally associated with the number 3. If, therefore, we abstract from all such classes their common property, we arrive at the class comprising all entities which have this property. So a useful definition of a cardinal number, e g 3, is that it is the class of classes which is equal to a given class (3 in this instance).

Now, no young child can be expected to know the word *class* in this sense, let alone define it; nor can he possibly define number as we have just done. But it is argued that the child through playing with pebbles, wooden bricks, etc, does build up for himself the concept of a class (e g all brown pebbles, all red bricks) and abstracts from similar classes (all containing the same number of elements) the concept of a cardinal number. That is, it is argued that logical concepts precede numerical or metrical ones. What is certain is that, when the child recognizes the quality of, say, 'fourness' in four pieces of furniture, four horses, four aeroplanes, he has built for himself the concept of a class without, perhaps, being aware of it.

It is not our intention to discuss these two viewpoints at length, and we shall limit ourselves to a very few issues.

First, we may agree with Meredith (1956) that modern mathematics may be *described*, though *not* defined, as the exploration of possibilities by means of logic. That is not easily understood by those who have had no training in mathematics, but it is a very important truth. We may say that mathematics begins with a set of undefined words: let us take the words *point*, *line* and *between* as examples. All other mathematical words are then defined in terms of such undefined words, together with a few words like 'and', 'the', which have no particular mathematical meaning. Again, mathematics can be said to start with a set of initial propositions called *axioms* or *postulates* whose truth is taken for granted; e g 'Through any two points there is one and only one straight line'. Such axioms are sometimes said to be 'self-evident', but that is not necessary. Provided that we have a set of meaningful statements that do not contradict one another, we can take any one of the statements and base further propositions or *theorems* on it. These are proved on the basis of the truth of the axiom and the laws of logic. So we can in one sense regard mathematics as a game where we make our rules and play the game accordingly. Hence in Whitehead's view there is now no option but to employ the term 'mathematics' in the general sense of being the science concerned with the logical deduction of consequences from the general premises of all reasoning.

Second, as Meredith (*op cit*) has also indicated, many mathematicians have resisted the thesis of the identity of mathematics and logic, since it conflicts with their personal, subjective experience. When discovering their theorems mathematicians seem to use intuition and insight, although they may well use logic in proving their theorems. By mathematical intuition is meant the insight one gains, without the intervention of conscious reasoning, into numerical, algebraic, geometric and other mathematical phenomena. Reasoning is used merely for verification of the intuition, although it must be made clear that intuition is dependent upon past experience and knowledge. Many mathematicians (c f Courant and Robbins 1958) would claim that constructive invention and intuition are at the heart of mathematical achievement and provide the driving force, although a neat logico-deductive system may well be a goal. Those who find it impossible to agree that all mathematics may be reduced to a system of conclusions drawn from definitions and axioms that must be consistent, would concede that the logical analysis of mathematics gives us a better understanding of mathematical facts and their relationships, and of the essence of the concepts involved. Again, mathematics was not analyzed in terms of logic until the present century, which suggests that it does have some pre-logical basis. Most teachers, too, maintain that at any stage in the teaching of mathematics, experience and intuition bring the first ideas to the child, and that, much later, the study of the logical relationships in which the ideas are embedded gives to the concepts a width and depth that were not obtainable earlier.

This discussion may seem very academic and far removed from the classroom. But it is not so in fact; it has been essential to raise the question of the logical basis of the natural numbers and mathematics generally before we discuss the varying approaches to the teaching of the subject.

In conclusion, it is worth noting that any real understanding of number seems to be the prerogative of man, in spite of the fact that some animals, e g birds, seem to be able to recognize up to about four objects or human beings. For example, a cat had three kittens, and if given a number of pieces of fish or meat she would eat all but three, leaving one piece for each kitten. One kitten died, and the mother cat thereafter left only two pieces. Now it can be said that the cat had some grasp of cardinal number. But only a fragmentary grasp. Our experience with other animals leads us to doubt if she could have matched six pieces of meat with six kittens and have made the adjustment if one died. Furthermore, there would never be in her mind, as far as humans can tell, any understanding that six comes between five and seven, or that it is equal to twice three, or equal to one half of twelve. Nor do we know if pussy would have recognized even the 'threeness' in three bottles, three children, three frosty mornings, and three loud noises. This example illustrates the fact that concepts are of all degrees of width and depth.

REFERENCES

COURANT, R and ROBBINS, H (1958). *What is Mathematics?* New York: Oxford University Press. 4th Edition.

MEREDITH, G P (1956). 'Mathematics and Mind'. *Mathematical Gazette*, **40**, 103–108.

WHEAT, H G W (1937). *The Psychology and Teaching of Arithmetic.* New York: Heath.

CHAPTER THREE

Some Approaches to Number Concepts—I

1 VERBAL METHODS

IT will not take long to discuss the verbal approach to the teaching of number concepts, since such an approach has long been regarded with disfavour in British Primary schools. In Great Britain verbal methods were *mechanical* in nature, rather than *formal*; the latter being more prevelant in continental schools.

In the mechanical approach the child may have acquired some idea of the natural numbers, through his experience of small groups in everyday life or of simple situations contrived in the classroom. Some concept of abstract numbers arose in the minds of many children whose teachers used this approach, but from the simplest situations involving small numbers there quickly followed much learning by rote, blind obedience to rules, the acquisition of 'tricks', and much oral and written practice. There is, of course, no objection to practice as such. Indeed it is necessary, as it helps to familiarize and fix concepts, and it gives the child confidence and skill in techniques. The great weaknesses of the method were that it did nothing to stimulate enquiry and discussion, it gave no place to imagination, and it provided no scope for the child to build up concepts by his own activity or experience. In essence, mathematical concepts were supposed to be built up mainly on spoken and written symbols, in the sense that the child, by manipulating these symbols, would come to comprehend the ideas underlying them. This view, that by repeating symbols verbally, or by manipulating them on paper, the child would come to grasp their significance, is a very old one. Augustine of Hippo in his *Confessions* said that 'One and one are two, two and two are four' was a weary chant for him. There is, of course, nothing wrong in employing some chanting to help a child to remember the addition table once he understands what it is he is manipulating.

On the Continent, where, in some countries, the age of entering school is later than in Great Britain, verbal methods employing a more formal approach have been used. The mathematics is based on definitions, which, as we have seen, are the formal bases. The work then develops as a result of successive deductions. Such an approach would be anathema to British teachers.

We must, of course admit that many able children made great progress in mathematics in spite of verbal methods; although it would generally be agreed that many children of below average ability made little advance, and even the able had less understanding of what they were doing than they would have had by using better methods. Some of the children who made good progress under such methods may have been those who had a stimulating environment outside the classroom, where there were opportunities for playing, arranging, ordering, grouping, discriminating, in very varying kinds of situations. This is only a possibility; we do not really know whether this was so in fact.

2 METHODS BASED MAINLY ON VISUAL PERCEPTION AND IMAGERY

The methods we are now going to discuss are frequently seen in use in Primary schools, very often combined with more 'active' methods that will be described later.

Let us begin by considering the concept of number itself. In essence the child is presented with a series of objects grouped together in space (they can be presented in sequence in time, but that is more difficult); later, pictures of objects may be used. In some way the perceptions, which are mainly visual, get transformed into knowledge. Other forms of perception may be brought in, as when the child touches the objects and experiences kinaesthetic and/or tactile perception. The child begins by playing with, say, beads, shells, etc and groups them according to a pattern or patterns, or he is presented with an arrangement of objects already grouped for him. He counts if necessary, says the appropriate number word associated with each group, learns to recognize the correct symbol representing the group, and later learns to write the symbols. After the use of actual objects, to afford further practice in recognizing groups, he may be given 'pattern cards' on which there are pictures of animals, objects, or dots. The elements forming the groups may be arranged in different patterns; for example, if dots are used:

The large dominoes used in infant schools are a good example of this kind of apparatus. Other examples are 'bead bars'— bars on which suitable objects can be threaded, the children having to thread on objects to a given number; also number strips, which consist of ten pieces of cardboard or hardboard of length 1 to 10 units, alternate unit lengths on the strips being coloured, say, black and white or red and blue. It will be remembered that Montessori sometimes used a similar approach. The shortest of her number rods was 1 decimetre long, and each of the other rods was increased by 1 decimetre, so that the longest was 1 metre in length. The sections were alternately red and blue. By *counting* the number of coloured sections the child was reckoned to arrive at a concept of 'how many'.

Some of the writers who have recommended the general approach we are now discussing believed that the concept of cardinal numbers (1, 2, 3, 4 etc) precedes that of ordinal numbers (1st, 2nd, 3rd, etc). They have suggested that the child does not at first see the group as a collection of equivalent units, but simply the total. The idea of ordinal numbers is then constructed on the basis of cardinal numbers, by arranging

34

the groups in ascending or descending order of size in such a way that the difference between any two adjacent cardinal numbers is the smallest possible. This is not the view of Piaget, as we shall see later.

Most of the authors who recommend methods included in this section seem to assume, although they may not be conscious of it, that the concept of number is built up by perception. It is reckoned that there develops a correspondence between the perceptual or physical structures of the number apparatus employed, and the mental structures which they evoke. Eventually, after ample experience of the physical structure, the mental structure remains in the mind even when the apparatus is no longer present. Thus the passage from the physical to the mental takes place, i e 'internalization' occurs. In the case of the Montessori or other material in which the layout of the apparatus, e g a pattern of dots, is rather stereotyped, the theory is that one kind of mental structure will develop for each number and that this will be roughly the same for all children. It has been pointed out that a possible drawback to all stereotyped presentation of groups (i e the elements always arranged in the same way) is the danger that the child will not recognize the number if the elements are rearranged. Indeed, this is often the case with young children. When the groups presented to the child are in flexible arrangements, and the child compares not only different patterns of the teaching material placed in front of him, but also notes the number of chairs, the number of children, the number of cups, and so forth, in the environment generally, presumably the varying impressions he receives of the groups give rise to differing images in his mind. But these images are linked by association in some way, and thus presumably the child abstracts the common feature of 'threeness' or 'fourness' or whatever the number may be, and so numbers become independent of the spatial relationships in which the materials are presented.

This general approach to teaching number, through visual perception and imagery, has been attacked on the grounds that it does not allow a child to see a number as a member of a system of relationships. While this was undoubtedly true in the case of much teaching using this approach, it is not always so. The child should, however, be brought very early to realize that a number stands in many relationships to other numbers; he should not be allowed to regard numbers as independent entities. A criticism of rather less weight is that the graphical representation of number often stopped at 10— a conventional rather than a natural limit. Another criticism is that this approach assumes a visual perceptual approach to be the best for every child possessed of normal sight. This criticism must be accepted until we know more about the ways in which children learn.

So far only the concept of number itself has been dealt with, but arithmetic operations could equally well have been taken to illustrate the approach. For example, in addition of number exercises, arrangements of dots, objects or pictures of animals might be used thus:

$$\begin{matrix} \bullet \ \bullet \\ \bullet \ \bullet \end{matrix} + \begin{matrix} \bullet \ \bullet \\ \bullet \ \bullet \end{matrix} = \, ?$$

Using number strips, the child would place, say, a four strip end on to a four strip and count up the unit lengths.

It is not, of course, suggested that teachers should use *either* methods employing visual perception and imagery *or* more 'active' methods. Indeed, in many infant and junior schools both kinds of methods are in use. Further, most approaches that depend mainly on visual perception and imagery, depend to some extent on actions as well.

A recent and thorough system of teaching arithmetic, based on this general approach is that employing the apparatus devised by Catherine Stern (Stern 1949). The distinguishing characteristic of the Stern material (the apparatus will be mentioned briefly a little later), is not that it can be manipulated, but that it should supply certain physical structures, which will lead to mental structures. The task of the child is to compare and measure the former. Stern claims that these structures will fix themselves in the mind of the child, just as graphic representations of numbers will. On page 21 of her book she points out that a child can see at a glance how two addends build up a

36

sum. Thus:

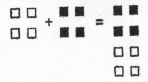

Figure 2

using her materials.

She continues, 'The structures of these patterns are unforgettable, so that the child can see the subgroups in his mind whenever he reconstructs the picture 8 or 9, etc'. Again, on page 289 we read, 'However, the materials have fulfilled their purpose only when the child can visualize them without their being present'. There is little doubt then that we are correct in placing her approach in this section. At the same time it must be stressed that the Stern material can to some extent be used to show the relationships between numbers *through the rearrangement* of the materials.

With this material the children are led to discover for themselves the answers to the problems presented to them; in this respect it is well in keeping with other modern methods. Furthermore, it is claimed that the child is presented, right from the start, with material that is so arranged as to give certain physical structures which will bring out the interrelationships between the numbers. Great use is made of coloured blocks of unit width and unit height, varying in length from 1 to 10 units and ruled into unit lengths. For example, each whole number between 1 and 10 inclusive is represented on a counting board by such a block. Hence we can demonstrate in concrete form such facts as that 5 is the number value of a group of 5 units, and that it contains one more unit than 4 and one less than 6. Moreover, since the ten blocks fit into a series of parallel grooves progressing from 1 to 10 units in length, the board shows that each number bears a relationship to all the others. There are also $\frac{3}{4}''$ hardwood cubes. Another piece of apparatus, known as a Number Track, is formed by runners or tracks separated by a space wide enough for the cubes to be

37

inserted. By joining together ten sections of the number track any number from 1 to 100 can be displayed. Again, there are squares of varying size that can be filled by four, nine, sixteen, twenty-five, etc, cubes.

It is claimed that insights into the following may be gained from the apparatus:

1 the meaning of counting,
2 the group and ordinal aspects of number,
3 knowledge of addition and subtraction facts within the range 1–100,
4 the meaning of multiplication and division,
5 operations involving units of measurement,
6 the nature of fractions and the concept of ratio,
7 the nature of decimals and the concept of percentage,
8 the nature of square measure and cubic measure.

A basic principle in the Stern techniques is that the child *measures* and *does not count*. It is argued that the piecemeal counting of single elements does not lead to an understanding of relations. But, using the number blocks (1) the ordinal number can be illustrated; (2) as can the cardinal number; (3) also the ratio meaning of number, in that a 7 block is 7 times as long as 1 block, can be shown; and (4) the child can grasp the relationships between numbers as he fits the blocks in various ways; for example, he can come to understand that the number 4 stands in a relationship to 1, 2, 3, 5; and that it can be related to other numbers to give 43, $\frac{1}{4}$, 4^3.

All the methods described in this section have this in common; they seek to achieve an intuitive cognition by presenting visual perceptual structures. A correspondence is then supposed to arise between the perceptual and physical structures, and the mental structures evolved. This is precisely what Gestalt Psychology would tell us, although many of the advocates of these methods would not have known this, for Gestalt theory postulates a correspondence between what we see and hear, and brain processes. There is no doubt about the influence of Gestalt psychology on Stern. Her book is dedicated to Max Wertheimer and she points out on page v that although she gathered her material empirically through teaching, it was

Wertheimer who provided the psychological background to her experiments. Later we shall see that, in Piaget's view, perceptual structures do not give the individual the manoeuverability of thought which is essential if mathematical concepts are to be fully developed. Indeed, there have been many critics of Gestalt theory in recent years.

Many teachers would employ the above type of approach in connection with other activities, such as shopping, weighing, measuring, centres of interest, and it is not intended to imply that the methods just discussed are always used in isolation. For example, Smith—formerly one of Her Majesty's Inspectors of Schools—in her book *Number*, recommends the following activities and pieces of apparatus: experience of groups from setting tables for mid morning milk (p 63); opportunities that arise in house play, baking, washing and other dramatic play (p 63); finger plays, nursery rhymes, stories and their extensions, such as drawing and modelling (p 64); bead frame (p 65); after groups have been built up in other ways, domino patterns (p 66); Tillich's bricks (p 68); make-believe play involving numbers (p 85); registration (p 87); shopkeeping (p 90); beads threaded on strings to illustrate tens and units (p 103).

But she does not favour the practice of hanging cards that have groups of dots with numbers underneath. Presumably, her experience led her to the opinion that in the early stages these cards were of little help. Yet, if the children had had experience of building up groups in other ways it is strange that she finds dominoes acceptable but not these cards. The use of both types of apparatus assumes that visual perceptual structures lead to mental structures.

REFERENCES

SMITH, T (1954). *Number*. Oxford: Blackwell.
STERN, C (1949). *Children Discover Arithmetic*. New York: Harper.

CHAPTER FOUR

Some Approaches to Number Concepts—II

ACTIVITY METHODS

'LEARNING BY DOING' has been a popular slogan in education during this century. It expresses the conviction that children's intellectual development takes place by their undergoing relevant activity rather than by remaining passive while being instructed by the teacher. John Dewey, the American educationist and philosopher, gave a great impetus to the spreading of this general point of view. His writings became well known from about 1890 onwards, although long before that date many teachers had recognized and written about the need for child activity.

Dewey's views on the psychology of number are given by McLellan and Dewey (1909). He believed that the idea of number is not impressed on the mind by the mere presentation of objects. The concept of number depends upon the way in which the mind deals with objects; in short, the mind has to compare and relate them in certain ways. This necessitates (a) the discrimination or recognition of objects as distinct individual units, and (b) generalization. Generalization in turn involves two sub processes:

i abstraction, or the rejection of all the qualities specific to each object save its 'oneness', and
ii gathering together objects to form a class or sum.

For Dewey, number can only be taught by presenting objects in such a way that discrimination, abstraction, and grouping are aided. There must be enough qualitative difference among the objects used for the child to be able to recognize the individual objects as distinct, but not difference enough to prevent them from being grouped to form a sum. Using suitable apparatus the child must classify, arrange in order, and compare. The resulting understanding of number comes from the actions involved, and not from the nature of the objects themselves, so that the concept of number develops independently of the physical structure of the teaching material employed.

Dewey rejects visual perception and imagery as bases of number concepts. Rather, the child's ideas of number are built up by using each number in many different situations that involve him in action. However, Dewey sheds no light on the way in which physical activity is transformed into mental activity. We must, however, mention his views briefly in passing, since he was a very influential figure among those who have tried to suggest a psychology of activity methods.

THE MODERN APPROACH IN THE INFANT SCHOOL

The approach to number in many British infant schools is one which may be termed the Environment Approach, since it seeks to use in the classroom the kind of experiences and situations that the child meets in real life. Kenwrick (1957 reprint) points out that experiences obtained by the child through projects and play give the motive, the reason for, and the meaning of number work. In number work, as in other departments of the work of the infant school, the great influence of educators such as Dewey and Froebel is immediately seen.

Children in the infants' school no longer engage as a class in number work, but in small groups, and at tasks that have often been chosen by themselves. One group of children is weighing dried peas, beans, flour, potatoes, sand, etc, while a second group may be weighing out sugar, fats, egg powder, etc, and making simple recipes. A third group may be measuring, with rulers, tape-measures, or yardsticks; and a fourth may be engaged in water measuring, with quart, pint and half-pint measures. In one corner there is a shop with much buying and selling taking place, while in the Wendy House further activities will be taking place, involving, say, counting or matching. The teachers' task is to provide a carefully-graded series of work cards containing problems that the child must solve in his activities.

Kenwrick (*op cit*) points out that after play experiences the child is ready for concrete material to enable him to discover principles for himself, e g that of place value. The time to introduce such apparatus, says Kenwrick, is when the need for it arises in the play situation. Furthermore, the child needs practice in working, say, the basic number bonds; and in connection with this certain games may be used. It requires skill on the part of the teacher to grade the work suitably and to ensure that each child is getting the maximum from his activities. Boyce (1953) gives an account of the kinds of activities that one might expect in the first year of an infant school using this approach, in the chapter entitled 'Growing to Understand Measurement'.

Those who advocate this approach do so on the following grounds. First, as the situations provided are real life ones, they are meaningful to the child. Second, because the child has freely chosen these activities, because they are purposeful to him, and because they are usually social and creative, he is likely to engage in them wholeheartedly and be completely absorbed by the task in hand. Under these circumstances he is most likely to learn rapidly, and to experience flashes of insight and understanding that he would not experience in a more formal learning situation, or using teaching materials that would control and direct his thoughts. In short, the child is reckoned to reach stages of thinking and insights that he would not reach in a situation planned for him.

This approach to number is in keeping with the general atmosphere of progressive infant schools today. The influence of Susan Isaacs is clearly seen at work, for in many schools, as at the Malting House School, children are encouraged to enquire for themselves, and an effort is made to bring within their experience every range of fact to which their interests reach out.

Isaacs would argue, as would Russian psychologists such as Vigotsky, that it is during play up to about 7 years of age that the foundations are laid for more complex forms of mental life. Few would disagree with this, but there is one matter over which we must be very careful when discussing the flashes of insight, or moments of understanding, that children get in unstructured (or structured) situations. Very often these flashes are unstable in that they occur in one situation or at one moment, but are as yet too fitful to form with others a system of thoughts that can be manoeuvred. The child's thinking is still very patchy and uncertain, and although he may well be able to manipulate a set of concepts (i e be capable of logical thought) in a simple and familiar situation, he has little or no power yet to merge one set of mental operations with another into a somewhat wider thought system. It is possible, if one is not careful, to overestimate the quality of the child's thinking. Nevertheless, it is from and by actions that the capacity to think at all comes.

Gardner (1950) has produced experimental evidence which, on the whole, is favourable to the approach practised in the progressive infant school. She compared children who had been taught along these lines in the infant school with control groups who had been taught by somewhat more formal methods, after all the children had reached the junior school stage.* A repeat of this experiment is now needed, employing large numbers of pupils and a follow up to fifteen years or later, to establish further, the value of the approach. The current N F E R research already mentioned in the preface should partly satisfy this need. Such an empirical study is necessary until we have a more comprehensive theory of child learning— one that will give a clearer understanding of the interplay between the affective and cognitive life. One of the great difficulties is that different children arrive at a knowledge of the same fact, or at the same stage of thinking, by what appear to be very different routes. Indeed, children get something out of almost every method and experience.

A doubt has arisen in the minds of some teachers of slow learners about the value of the emphasis on non-directed activity. They argue that while such activities may be of the greatest value to ordinary children and fast learners, in dealing with

* For a short account of Gardner's findings alongside those of Jersild, who carried out a much larger experiment in America, see Rusk, R R, *An Outline of Experimental Education*. London: Macmillan, 1960, ch 8.

43

slow learners the teacher must spend rather more time in directing the attention of the child to some activity, and must try to stimulate him into thinking. Experimental evidence on this issue is also required.

FURTHER DEVELOPMENTS IN ACTIVITY METHODS

Piaget's views, which will be further elaborated later, are that the child does not make abstractions directly from the handling and manipulation of materials. Rather, abstraction arises through the child coming to appreciate the significance of the transformations that take place as he classifies objects, and puts them in order of size; and as the objects are rearranged first to yield one perceptual structure then another, changed from one situation to another, and so forth. Piaget believes that mathematical concepts are not derived from the materials themselves, but from an appreciation of the significance of the operations performed with the materials. The concepts and the ability to manoeuvre them in the mind, he considers, are built up from using concrete material, but are independent of the actual materials used. When the child has come to appreciate the significance of his actions, and says, for example, 'I am placing the sticks longest to shortest', or 'I am putting all the wooden beads together', he is likely thereafter to be able to perform certain manoeuvres in his mind in relation to these activities. But mere perception and imagery in relation to the materials do not necessarily permit these mental operations, i e true thinking. This point can be illustrated as follows. Suppose a child argues, 'All boys (A) plus all girls (B) equals all children (C)'. This is a mental operation or a thought, and in Piaget's view, it is flexible and reversible, so that the child could also think, $A = C - B$, $B = C - A$, $C - A - B = 0$. A true intellectual operation allows a variety of compositions; but when a perceptual field is restructured or reorganized it is nearly always in one direction only, and leads from one rigid Gestalt or pattern to another. This reorganization of the perceptual field may be due to a change in the location of the viewer, or in the exact point on which he centres his attention. Perceptual reconstructions are thus not generally capable of being broken down and reassembled in a different order;

44

the representations that the child makes to himself by means of imagery cannot be combined with others and reversed. In Piaget's view, then, Gestalt theory neglected the fact that perceptual structures are fairly rigid, whereas mental operations are flexible and can be rearranged in many different ways.

Number apparatus has been devised, however, which, it seems, meets some of these objections, in that:

i   it clearly enables the child to appreciate the significance of his own actions through rearrangement of the materials,

ii it yields concepts which are mathematically valuable,

iii it relies only in part upon visual perception and imagery.

A carefully thought out set of materials fulfilling these conditions has been devised by Georges Cuisenaire, a Belgian.

THE CUISENAIRE APPARATUS

The materials and their use have been described by Cuisenaire and Gattegno (1954), also Gattegno (1958, 1960), and Gattegno (1957) has published a set of arithmetic textbooks for primary school children in which the exercises are based on the use of the materials. Clarkson (1959) has given a short report on the use of the Cuisenaire material as it has been used in the infant school of which she is headmistress.

The material consists of a number of rods of ten different colours, varying in length from 1 cm to 10 cm. Each is 1 cm by 1 cm in cross section. All rods of the same length are of the same colour. The apparatus is used in two stages. During the first the child:

i   plays freely with the rods,

ii associates a colour with a length (though *not* with a number name at first),

iii puts two rods end to end and says, for example, 'The light green rod plus the crimson rod equals the black rod'. Covers up part of one rod with a shorter rod and says, for example, 'Blue rod minus a yellow rod equals a crimson rod',

iv places, say, three red rods end to end and says, 'Three red rods equal one dark green rod',

45

v   rods can be found such that two or three of one equal one of the other.   Child says appropriately, 'The yellow rod is equal to half the orange rod', etc.

During this stage the child has also learnt to associate a symbol (e g 6) with a group of shells, chairs, etc, and to write the symbol. He has also learnt the meaning of 'plus', 'minus', 'equals', etc, and can write the correct signs for these words.

At the second stage numbers have to be linked with a length. The teacher may begin by asking, if the white rod (1 cm long) is 1, what the lengths of the other rods are. By using the correct rods the child is then ready to record these operations:

$$4+3=7 : 3+4 = 7 : 7-4 = ? : 7 - ? = 4 \text{ etc.}$$

So the addition and subtraction operations are understood and facts learnt. By means of these rods, a wall chart, and some cards, it is also possible to teach multiplication and division. Later, the material can be used to give an understanding of fractions and proportions, and it can also be used to demonstrate operations involving money and some other weights and measures.

The Cuisenaire material certainly makes some use of imagery as well. For example, Gattegno (1958) p 29 writes, 'Our stress on dynamic thinking, from the situation to the image, and from first images to more elaborate ones will considerably simplify teaching . . .' Moreover, the concepts of number and number operations that the child forms are due to what he does as well as what he sees.   However, neither Cuisenaire nor Gattegno has provided a convincing theory linking the interplay of perception and actions, or of the relation between them and the mental structures that result therefrom, although in his more recent writings Gattegno increasingly adopted a Piagetian viewpoint. It will be noted that in using these materials there is no counting, since the rods are not marked off into lengths, and the child has to remember the length of each and the colour. But the apparatus does present a structured situation to the child in which it is comparatively easy for him to discover for himself many mathematical relationships.

It is unlikely that any school uses the Cuisenaire or other apparatus to the exclusion of all other approaches; an environmental approach is likely to be used at the same time. The Cuisenaire apparatus, like that of Stern, helps to produce the mathematical concepts and the knowledge of operations needed in solving the problems the environmental material provides.

THE VIEWS OF PIAGET

A few experiments performed by Piaget (1952) will be described. In one, about twenty beads are placed in a box. Most of the beads are brown, the rest are white; all are made of wood. The child is asked if there are more brown beads or more wooden beads. To answer this the child has to consider the same beads twice in the process of making his comparison, since the brown beads are also made of wood. Piaget maintains that up to 7 years of age the child nearly always says that there are more brown beads than wooden ones, because there are many brown beads but only one or two white ones. The child's problem seems to be this. While he is centring his attention on the class of brown beads, he loses sight of the total class of wooden beads, and so he can only compare the class of brown beads with the complementary class of white beads. The child's thinking, in Piaget's view, is too much influenced by his perceptions, which can be misleading. At first, they do not give the child a grasp of extensive quantities, that is, of the relationship between the part and the whole as well as between the parts themselves. Indeed, at first his perceptions lead him to mix extensive and intensive quantities (the latter indicating the connection between the part and the whole without taking into account the relations between the parts), in such a way that he cannot differentiate between the two, and does not comprehend the idea of a group (here a total group of wooden beads). For Piaget, perceptions and associated images will not provide the concept of a group, because as we have seen, they are rigid, irreversible, and cannot be rearranged in various ways. Later, the child's thought will be more manoeuverable or operational; he will be able to think, 'All brown beads plus all white beads equals all wooden beads', 'All wooden beads less all white beads equals all brown beads'.

Hyde (1959) in a study of 144 children in Aden,* aged 5–9 years, repeated this experiment, and confirmed Piaget's results. Nevertheless, she adds a note of caution. Although the sequence of replies was much as Piaget suggests, she doubts if the child views the problem as Piaget believes a child does. Hyde suggests that mental set or expectation enters into the results. The moment the child sees the beads he notes that there are a very few of one colour and a far greater number of another colour. For the less wary child (we do not know how many fall into this category) the heart of the problem is 'What has talk of wood got to do with it?' This seems to be a good example of set, expectancy or attitude adversely affecting thinking.

Experiments carried out under the writer's direction among primary school and E S N special school pupils, into children's ability to consider the whole part relationship, have confirmed Piaget's findings in the main, although among the E S N pupils the ability does not develop until secondary school age. However in all children, the whole part relationship can be grasped more easily in some media than in others. In our work it was found to be more readily understood using coloured beads than it was using different, but brightly coloured flowers.

Piaget suggests that by 6 years of age the child may have some intuition of numbers up to six in much the same way as animals have a knowledge of small numbers. Furthermore, he may be able to count, but this does not imply that he has much concept of number. This was illustrated by the following type of experiment. The child made a row of, say, five counters on a table. He then laid out a parallel row containing the same number of counters by establishing a one-to-one correspondence, and would admit that the two rows contained the same number of counters. But when the counters in one row were spread out so that the perceptual correspondence was lost, the child up to 6–7 years of age would no longer agree that the two rows now contained the same number of counters. Once again it seems that his perception has led him into error; he cannot yet think, in effect, that if one row is longer, the counters in it are further apart. When he can, he is able to

* There were 48 English, 48 Arab, 24 Indian and 24 Somali children.

perform the actions in his mind without actually performing with materials; that is, his thinking has become operational, and it is no longer necessary for the counters in each row to be opposite each other for him to grasp that the number of counters is the same in both rows. His thinking is now at the concrete operational stage; that is, he can 'think' in relation to real situations. Furthermore, if now the rows are made up of different objects, he continues to realise that the number of objects in each row remains constant. Hence Piaget maintains that for the child to be able to establish arbitrary correspondence in these varying situations he must have attained the ability to form the concept of a class, from which it follows that the latter ability underpins the development of the concept of number.

Hyde (*op cit*), in repeating this type of experiment, used small rubber dolls, which had to match with toy tin baths. Six baths were placed side by side as close as possible, and the child was given a tin of dolls and told to put a doll in each. He was questioned as to whether there was the same number of baths and dolls. The child was then told to put the dolls in a heap and questioned again. Finally the dolls were spaced out evenly in front of the baths, but displaced to one side of the

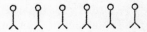

Figure 3

baths as in Figure 3, and the child was requestioned. In general Hyde confirmed the stages given by Piaget, but notes that it was a difficult test for the multiracial group of children aged 5-9 years, although even the youngest could count beyond six. Ability to count is therefore no guarantee that the child has a stable concept of number.

Beard (1957) found that about three-fifths of her group of 6-7-year-old children could match a line of twelve counters with similar counters.

Another mental operation is necessary before the number concept can be formed—namely, the ability to arrange a series of items according to their differences. Up to about 5 years of age the child is unable to make a series of, say, sticks of increasing length; between 5 and 6 he arrives at the seriation* by trial and error, but if a stick is omitted he has difficulty in inserting it into the correct position in the series. For Piaget this means that the child cannot decompose the series mentally. But by seven the child can take the shortest stick, then the next shortest, and so on, knowing that he will obtain a series of sticks of increasing length. He has now reached the stage where he can co-ordinate in his mind the two relations, 'stick A is shorter than stick B, and stick B is shorter than stick C', and he can mentally pass through the series in both directions.

Piaget also maintains that when the child can seriate and establish correspondence in the mind (i e not by trial and error), he can find the cardinal number of the class (say 7) preceding a number defined by its ordinal position (8th). For example, if the sticks are 'stairs', he can say how many 'stairs' a doll has climbed when it reached a certain step on the 'staircase'. He comes to understand the cardinal and ordinal meanings of number at the same time.

At this point Hyde (*op cit*) repeated two experiments to check Piaget's views. First, a number of sticks of identical material and cross section, but differing in length, had to be placed in order of length. The smallest was 9 cm long and they increased in length by 0·8 cm intervals. When they had been placed in order the child was handed a second set of sticks, the shortest of which was 9·4 cm, the remainder increasing in length by 0·8 cm intervals. He was told to put these in their 'right places' and so make a staircase.

Second, Hyde used ten cardboard strips each 1 inch wide, the first being 1 inch high, and the height of each succeeding strip increasing by 1 inch. The child had to make a staircase of the strips beginning with the smallest. The experimenter would select a strip and say, 'How many strips like this (smallest) could we make from that strip?' By counting along the bottom edge of the strips and getting the position of the

---

* Seriation is the formation of, or the putting into, a series.

given strip (ordinal number) the child could tell how many of the smallest strips could be made from the given strip (cardinal number).

Beard (*op cit*) conducted similar experiments. For example, the child had to say how many 'stairs' (cardinal number) a doll would descend in walking to the bottom from a given stair (ordinal number). She also had strips of differing lengths, as in Hyde's second experiment, and similar kinds of questions were asked.

The conclusion to be drawn from these experiments is that the stages proposed by Piaget are broadly confirmed, although they do not necessarily confirm that children arrive at an understanding of the cardinal and ordinal aspects of number at the same time. For example, Hyde specifically states that quite a few children who failed to put the strips in serial order could say how many little strips would be needed to make a bigger strip.

Summing up so far, we can say that for Piaget the concept of number is not based on images or on mere ability to use symbols verbally, but on the formation and systemization in the mind of two operations; classification and seriation. For the concept to form in the mind these two operations must blend, and in order that objects may be both equivalent and yet different, the qualities that are specific to each member of the group must be eliminated so that the homogeneous unit 1 may be formed; for instance the characteristics that distinguish, chairs, jugs, and shells are eliminated and each is regarded as a thing. A practical example will help to make the general position clearer. As the child picks out all blue beads and puts them in a box he comes to think of all the blue beads together and eventually forms the concept of 'class of all blue objects'. By sorting other materials he can likewise form the concept of other classes; the concept of a class, or the mental operation of classifying, is an internalized version of grouping together objects as similar. Again, in his activities involving seriation he puts bricks in order of length from, say, left to right. From this kind of action he derives the concept of relations. The number system is the union of classification and ordering, for the idea of the number 8, say, depends upon the child grouping in his mind eight objects to form a class, and upon placing 8 between 7 and 9; that is, in relation.

In the experiments described on page 46 the child could not manipulate in his mind, the relationships between (a) the brown and white beads and (b) the whole collection. That is neatly illustrated again in another experiment. Children were given four beans to represent four sweets, and were told that these were to be eaten during the morning break. Four more 'sweets' were to be eaten at tea time. They were then told that next day they would receive the same number of 'sweets' again and two groups of four were set in front of them. But, on this occasion, only one was to be eaten in the morning, the remaining three were to be eaten along with the other four 'sweets' at tea time, and these three 'sweets' were placed with the four, the children watching the procedure. They were then asked to compare the two sets in front of them, 4+4 and 1+7, and asked if they would eat the same number of sweets on each day. The 5–6-year-olds said that 1+7 was either smaller or larger than 4+4, their answer depending on whether they compared the 1 or the 7 with the 4's. But by 7–7½ years of age the correct answer came immediately and further questioning showed that the children understood that the differences were compensatory. At this stage the 'sweets' must be seen as a collection of units not identical, yet capable of being equalized, units serving as a common measure. Such an experiment shows how children could learn the addition or subtraction table, and repeat it to the teacher's satisfaction and yet have little or no idea what they were saying.

Hyde (op cit), using piles of shells, found very much the same as Piaget, although Beard (op cit), using sweets, found the children used more varied methods of comparing 1+7 with 4+4 than Piaget records.

For Piaget, logical concepts precede numerical ones. Mathematical concepts cannot be brought about by using the symbols of mathematics, by verbalizations, by mechanical processes, or by perceptual reconstructions. They are arrived at by manipulating things but do not come from the things themselves. On this view children should be given materials that can be made up into different collections according to different criteria;

children should establish correspondence, order, include one class within a more general class, etc. For example, a child should establish correspondence between his row of shells and that of his comrade. Churchill's work (*op cit*) suggests that suitable play activities may help forward the development of number concepts. Such activities can be undertaken in the infant school.

A number of references have been made to the work of Beard (1957) and of Hyde (1959), and it appears from these references that their work often confirms the findings of Piaget. That is true within limits, but one important finding not in agreement with his general viewpoint is that a child can be at a concrete operational stage of thinking in one test but not in another. That is also confirmed by Williams (1958) who found that in tests of cardination and ordination, etc, children who on some tests were clearly at the concrete operational stage were at the pre-operational stage on others. Again, some children who could appreciate ordinal positions appeared unable to appreciate the significance of cardinal groupings. Similarly Mannix (1960) has shown that E S N Special School children are not necessarily at the same stage of development from test to test, on a number of Piaget-type tests involving number concepts. Again, Dodwell (1960, 1961) gave a number of Piaget's tests to 250 Canadian primary school children and confirmed the general kinds of behaviour described by Piaget. But he also found inconsistency of type of response from test to test, and variability from child to child at a particular age in the sort of response made. An article by Hood (1962) is very interesting. He has shown among British primary school children that although the majority of children who could satisfactorily pass the Piaget tests, of the type just described, were making good progress in their arithmetic, some 10–15 per cent could perform the tests but were clearly backward in their school number work. The present writer has also found that secondary school children who are very backward in number can often pass the tests. These workers have thrown some doubt on the value of Piaget's theories in their entirety, but they in no way detract from the value of his ingenious experiments. It is likely that children arrive at these concepts partly

53

as a result of logical reasoning and partly as a result of sheer experience in the manipulation of objects.

An interesting study by Wohlwill (1960) shows three distinct phases through which children pass in understanding number. First, they respond to it on a perceptual basis so that if the arrangements of the elements is changed the number of elements may change. Second, numbers are responded to conceptually. In passing from the first to the second phase mastery of the correct verbal terms for the groups plays a prominent role. Third, the relationships among individual numbers are conceptualized.

In a later paper Wohlwill and Lowe (1962) show how difficult it is to provide learning situations for young children in order to speed up their understanding of the principle of number conservation. This finding agrees with that of Smedslund (1961) who tried to speed up the learning of the conservation of weight (Chapter 6). Such learning was found to be limited, and the understanding of conservation compared unfavourably with the child's understanding when conservation appeared spontaneously.

THE DIENES MATERIALS

Dienes (1959, 1960) has given some details of materials which have been in use in Leicestershire junior schools, and which, he claims, help in aiding children to develop mathematical concepts. At the outset he makes a clear distinction between *analytic* and *constructive* thinking as we have already noted. In the former kind of thinking there is an awareness of all possible logical relationships, so that concepts are formulated explicitly and exactly before they are used. On the other hand, in constructive thinking, concepts are built out of some broad general picture, without the child being aware of all possible relationships. Dienes makes clear that these are not necessarily the only kinds of thinking, nor are they mutually exclusive, and holds that for most people constructive thinking develops before analytic. In the school room, however, children are likely to be taught concepts (if this can be done!) according to the way the teacher thinks they develop in his pupils.

Dienes posits three hypotheses which he thinks may have an

a priori face validity, and which have affected the design of his apparatus. They are:

1 The greater the degree of generality at which the concept is formed, the wider is the field in which it may be applied.

2 If a concept involves a number of variables, to ensure general understanding all the variables must be varied.

3 A concept is achieved more rapidly if the individual receives many different perceptual impressions (visual, tactile, kinaesthetic) that illustrate it.

He also classifies mathematical concepts, as far as their content is concerned, into three types:

(*a*) pure mathematical concepts. These deal with numbers and the relationship between numbers and are independent of the way in which numbers are expressed; that is, independent of the Roman, Hindu-Arabic or other notation,

(*b*) notational concepts. These deal with the properties that arise because of the way in which we express numbers; e g the concept of 'place value' in the Hindu-Arabic notation system,

(*c*) applied concepts. These include length, weight, time, etc and, when any one is considered, a specific facet of reality is being studied to the temporary exclusion of other facets.

A considerable amount of apparatus has been devised to aid the development of all three kinds of concepts. For example, under (*a*), there is apparatus to help the formation of the mathematical concept of square (not the geometric concept of square). After using pieces of apparatus involving square and rectangular shapes the child is presented with a number of small equilateral triangles, the side of which is regarded as unity. These small triangles have to be placed along one side of an equilateral triangle of side x; further, the child discovers that to fill in all the large triangle x^2 triangles are required. He is then told to add another row of small triangles along one side of the large triangle (this row is indicated by dotted lines). The number of small triangles is now $(x+1)^2$, which is equal

Figure 4

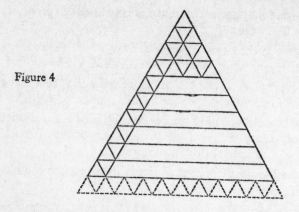

to $x^2 + 2x + 1$. Thus another step is taken towards the general case $(x+y)^2$. By this means Dienes claims that high order concepts can be made available to younger and less able pupils than if using more conventional methods. Further, this constructive approach does no harm to the child who can think analytically, for he will discard the apparatus if it is no longer required.

By means of what are called Multibase Arithmetic Blocks children can be helped to develop notational concepts, e g place value. Pieces of wood are marked off into unit cubes as in figure 5.

The ratio may be 3, as illustrated, or 4, 5, 6, 10, etc. These pieces are referred to as *units*, *longs*, *flats* and *blocks*. By means of them we can add or subtract using a base of 3, 4, 5, 6, or 10 as we normally do since, say, 3 units = 1 long, 3 longs = 1 flat, 3 flats = 1 block. This kind of apparatus, too, should prevent confusion between $2x$ and x^2, since two *longs* do not look in the least like a *flat*.

Dienes makes the point that his Multibase Arithmetic Blocks are more useful in helping the pupil to develop concepts involving indices and logarithms than either the Stern or Cuisenaire materials. For the blocks enable both the base and index (power) to be varied, whereas the Stern apparatus enables the power to be varied but makes no provision for a change in base. Neither is built into the Cuisenaire material (Dienes 1959 p 24).

The use of the Dienes materials ensures that children are

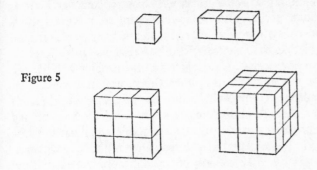

Figure 5

put to work with apparatus that gives a structured situation in which it is easy for them to find for themselves certain mathematical relationships. Just how far these relationships will be generalized we do not know, since it depends to a great extent upon the child and his capacity for transferring thinking skills from one situation to another. It will be seen that Dienes's views are in contrast to those of Piaget over certain psychological issues, since he maintains that the child is more likely at first to build concepts intuitively without being aware of relationships between these and other concepts. Nevertheless, using these materials, the child seems to arrive at certain concepts partly as a result of perceptual structures, and partly as a result of coming to appreciate the significance of his actions.

CONCLUSION

Some people, though not wishing to decry the creative spontaneity of children, doubt if the environmental approach, used exclusively, can be adapted to bring out essential mathematical relationships. They suggest that some kind of structured approach should be used to help children to develop these relationships, which *should then be applied to solve the kind of problems found in the environmental approach*. The June, 1959, issue of the *New Era* was devoted to a discussion of this problem, and should be read. Keen advocates of the environmental approach may deny that any kind of structured material should be used before 7 years of age, but may be in favour of their use thereafter.

It is possible that some children get more from one kind of structured material, and some children from another. A recent survey of the studies made into the effectiveness of these varied materials has been provided by Williams (1965). In the same paper Williams also deals with the problems of design in testing the effectiveness of methods and materials. To date, experiments involving the use of different kinds of materials have been on too small a scale, and have been carried on over too short a period for indisputable conclusions to be drawn from them. Hence the importance of the investigations into the relative effectiveness of various approaches that are now being carried out by the National Foundation for Educational Research.

All pieces of apparatus used to illustrate mathematical principles, excluding geometrical models, are concrete 'analogues'. They are something the same and yet entirely different from that which they represent (c f Williams, 1961, 1962, 1963). At the same time, some analogues may aid concept-formation more than others.

It is certain that much thought is currently being given to both teaching methodology and changes in the mathematics curriculum in British primary schools. In the summer of 1965 the Schools Council published *Mathematics in Primary Schools* (London: HMSO), while the first draft of some of the books containing suggestions from the Nuffield Mathematics Teaching Project (*I Do and I Understand*, *Computation and Structure*, *Size and Shape*) became available. At the same time, however, many problems remain regarding mathematics learning, and readers should consult the work edited by Dienes (1965).

In Chapter I it was stated that from about 7 to 8 years of age onwards the child can 'turn round on his schemas'. He is then *aware* of the sequences of action in his mind, he can see the part played by himself in ordering his experience, and for any action in his mind he can see that there are other actions that will give the same result; that is, he sees equivalences. Thus he understands $5+2=7=6+1=9-2$. This new co-ordination of schemas enables the child to measure the same distance in feet or in inches, perform subtraction using the method of complementary addition, and spontaneously decompose 35 into tens and units or sixes and units depending upon the number base that is being used. There is now learning with understanding (cf Williams, 1964). It will be seen that getting the child to appreciate the significance of his own actions, so that he becomes aware of what it is he is doing, is crucial. Unfortunately we still do not possess the knowledge that will enable us to ensure that this understanding will take place.

REFERENCES

BEARD, R (1957). 'An Investigation of Concept Formation among Infant School Children'. Unpublished PhD Thesis: University of London.

BOYCE, E R (1953). *The First Year in School.* London: Nisbet.

CLARKSON, J (1959). 'Using Cuisenaire with Infants'. *New Era*, June.

CUISENAIRE, G and GATTEGNO, C (1954). *Numbers in Colour.* London: Heinemann.

DIENES, Z P (1959). 'The Growth of Mathematical Concepts in Children through Experience'. *Educ Research*, **2**, 9–28.

DIENES, Z P (1960). *Building up Mathematics.* London: Hutchinson.

DIENES, Z P (1965). *Current Work on the Problems of Mathematics. Learning.* Hamburg: UNESCO.

DODWELL, P C (1960). 'Children's Understanding of Number and Related Concepts'. *Canad J Psychol*, **14**, 191–205.

DODWELL, P C (1961). 'Children's Understanding of Number Concepts'. *Canad J Psychol*, **15**, 29–36.

DODWELL, P C (1962). 'Relations between the Understanding of the Logic of Classes and of Cardinal Number in Children'. *Canad J Psychol*, **16**, 152–160.

GARDNER, D E M (1950). *Long Term Results of Infant School Methods* London: Methuen.

GATTEGNO, C (1957). *Arithmetic. Introductory Stage. Books 1, 2 and 3.* London: Heinemann.

GATTEGNO, C (1958). *From Actions to Operations.* Reading: Cuisenaire Co.

GATTEGNO, C (1960). *A Teacher's Introduction to the Cuisenaire-Gattegno Method of Teaching Arithmetic.* Reading: Gattegno-Pollock Co.

HOOD, H B (1962). 'An Experimental Study of Piaget's Theory of the Development of Number in Children'. *Brit J Psychol*, **53**, 273–286.

HYDE, D M (1959). 'An Investigation of Piaget's Theories of the Development of the Concept of Number'. Unpublished PhD Thesis: University of London.

KENWRICK, E E (1957). *Number in the Nursery and Infant School.* London: Routledge and Kegan Paul.

MANNIX, J B (1960). 'The Number Concepts of a Group of E S N Children'. *Brit J Educ Psychol*, 30, 180–181.

McLELLAN, J A and DEWEY, J (1909). *The Psychology of Number.* New York: Appleton.

PIAGET, J (1952). *The Child's Conception of Number.* London: Routledge and Kegan Paul.

SMEDSLUND, J (1961). 'The Acquisition of Conservation of Substance and Weight in Children'. *Scand J Psychol*, 2, 11–20; 2, 71–84; 2, 85–87; 2, 153–155; 2, 156–160.

WALLACH, L and SPROTT, R L (1964). 'Inducing Number Conservation in Children'. *Child Develpm*, 35, 1057–72.

WILLIAMS, A A (1958). 'Number Readiness'. *Educ. Review*, II, 31–45.

WILLIAMS, J D (1961). 'Teaching Arithmetic by Concrete Analogy'. *Educational Research*, 3, 112–125.

WILLIAMS J D (1962). 'Teaching Arithmetic by Concrete Analogy—II'. *Educational Research*, 4, 163–192.

WILLIAMS, J D (1963). 'Teaching Arithmetic by Concrete Analogy—III'. *Educational Research*, 5, 120–131.

WILLIAMS, J D (1964). 'Understanding and Arithmetic: Some Remarks on the Nature of Understanding'. *Educational Research*, 7, 15–36.

WILLIAMS, J D (1965). 'Teaching Methods, Research Background and Design of Experiment'. London: NFER. *Arithmetic Research Bulletins*, No 1.

WOHLWILL, J F (1960). 'A Study of the Development of the Number Concept by Scalogram Analysis'. *J Genet Psychol*, 97, 345–377.

WOHLWILL, J F and LOWE, R C (1962). 'Experimental Analysis of the Development of the Concept of Number'. *Child Develpm*, 33, 153–167.

CHAPTER FIVE

The Concept of Substance

By the word 'substance' is meant 'amount of matter',*
'amount of material', or to use the term of the physicist—'mass'.
It is intended here to discuss how the child's ideas of 'amount
of material' develop with age and experience.

Piaget (1950, 1953a and 1955) has shown how, in his view,
the child slowly builds up the idea of the 'object' during the
first two years of life. During this period he gradually comes
to distinguish between his own body and the other objects of
the environment, and builds up a picture of the world as
consisting of a number of objects that continue to exist even
when they disappear from his sight; and he learns that, as a
rule, they maintain their shape and size. He also comes to
regard himself as one person among the others around him.

It is not, however, intended to consider further the development of the child's concept of the object at this early age. It is
sufficient to say that by the second birthday the ordinary child
apprehends a cup of milk, a piece of plasticine, a brick, a doll
etc as independent entities, and that they remain the same as
he looks at them; and that, if they pass from his gaze momen-
tarily, he will find them as he left them, provided they have not
been disturbed in the interval. For the adult such a statement
is trite, but for the child it indicates real discoveries.

The child's play with sand, water, and the like, whether in the
home environment, or nursery and infant school, is of the
greatest value to him in helping him to understand that he has
a certain quantity of stuff in, say, his hands, or in a tin. Of

* Equally well 'quantity of matter'.

course, maturation as well as experience determines the child's ability to understand. He slowly builds up a vocabulary, and, after a period during which he may use words without having much idea of their significance, he comes to understand that he has 'much', 'little', 'a lot of', 'the same' and so forth, although he cannot, of course, at this stage use terms such as 'amount of matter'. Furthermore, if he adds sand to the heap he already has, he comes to understand that he has 'more', and if he throws away some of his water he comes to know that he has 'less'. As far as we can judge the child comes to realize, as a result of age and experience, that the amount of milk in a cup, the quantity of sand in the pit, the lump of plasticine that he sees on the table, remains the same, at least for very short periods, if nothing is added or taken away, and if the substance is not disturbed in any way.

PIAGET'S VIEWS ON THE CHILD'S CONSERVATION OF SUBSTANCE

Only in recent years, however, have we found out that up to 7–8 years of age or so, the average child does not appear to understand that the amount or quantity of matter stays the same regardless of any changes in shape or position. That is an illuminating piece of insight into child thinking for which we have to thank Piaget in the first instance. This concept of conservation of substance (or invariance of substance) is an important one, for the mind can only deal effectively with a lump of plasticine, a glass of water, or a collection of shells, if they remain permanent in amount and independent of the rearrangement of their individual parts.

As was stated when dealing with number, Piaget's work suggests that a child's thinking is largely influenced by his perceptions between 4 and 7 years of age. During this period it seems to be so determined because he centres his attention on one aspect, dimension, or element of the situation and ignores other aspects. But from 7 to 8 years of age onwards he is increasingly able to break away from the influence of his perceptions and, in Piaget's view, gains in power to apply logical thought to practical problems and concrete situations. As we saw earlier, he is supposed to come to appreciate the significance of his own actions, so that they acquire the form of reversible operations in the mind and thus render him less dependent upon perception.

From his many experiments involving the conservation of continuous substances (e g plasticine, water) and discontinuous substances (e g beads, shells), Piaget concluded that children pass through three stages; namely, non-conservation, transition, and conservation. In a typical experiment a child was given two equal balls of modelling clay, which he (the child) agreed were of the same size. One of the balls was then rolled out to look like a sausage. At the first stage the child denied that the amounts of clay in the ball and sausage were the same. He may have made a response such as, 'It's more because the sausage is longer'. At the transition stage, he arrives at the concept of conservation under some conditions, or at one moment, but will lose the idea under slightly changed conditions. But according to Piaget, from 7 to 8 years of age he increasingly feels the logical necessity for conservation and will support it by argument. For example, Piaget (1950 p 140, also 1953b p 16) maintains that the child says the sausage can be returned to the shape of a ball, or what has been lost in one direction has been gained in another, or that nothing has been added or taken away.

In another typical experiment two identical vessels A and B are filled to the same height with a coloured liquid. Young children of 4–6 years of age will admit that the amounts of liquid in the two vessels are the same. Next the liquid in vessel A is left undisturbed, but the liquid in B is poured into

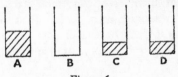

Figure 6

the vessels C and D, which are identical in size and shape to A and B, so that the amounts of liquid in C and D are the same (Figure 6). The child is then asked, in suitable language, if the amount of liquid in C and in D together is the same as the amount of liquid in A. Children of 4–6 years of age deny that the amounts are still the same. They notice that the levels are lower, and so, for them, there must be less liquid. Or, they notice that there are two vessels, and so consider that there must be more liquid.

Even at 6–7 years of age some children do not affirm that the amounts of liquid are the same. If the liquid in B is poured into the two vessels C and D they may agree; but if it is poured into a wide vessel (E) so that the depth of liquid is very small, or into a thin vessel (F) so that the column of liquid is tall (see Figure 7), they will deny conservation. It seems that it is

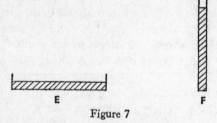

E   F

Figure 7

not a lack of verbal understanding that causes the child's confusion since he admits to conservation when perceptual differences are small, but denies conservation when perceptual differences are great.

By 7–8 years of age, the child admits to conservation and stands firm in his conviction; and, as already stated, Piaget maintains that the child can only arrive at the concept because logical concepts are already at his disposal. He claims that the child can reverse the process in his mind, thus employing the logical concept of an inverse operation; or he can grasp that what has been lost in one dimension has been gained in another, so employing the logical concept of compensation or displacement of parts ('Taking from here and putting there').

FURTHER EXPERIMENTAL EVIDENCE ON THE CHILD'S CONSERVATION OF SUBSTANCE

Lovell and Ogilvie (1960) report work in which 322 children in a north of England junior school were examined individually in order to study the growth of the concept of conservation of substance. Balls of plasticine about two inches in diameter were used. Providing the child agreed that the amounts of plasticine in the two balls were equal to begin with, he was deemed to be a suitable subject for the experiment. One ball was rolled out to form a 'sausage', and the subjects were questioned at some length regarding the amounts of material in the 'sausage' and ball. The accompanying Table 1 shows the percentage number of children found at each of the three stages mentioned above, for each of the four years of the junior school (7 + to 11 + years); the number of children in the year groups being respectively 83, 65, 99, 75, from youngest to oldest.

TABLE 1

No of Children	Stage 3 Conservation %	Stage 2 Transition %	Stage 1 Non-Conservation %
1st year (83)	36	33	31
2nd year (65)	68	12	20
3rd year (99)	74	15	11
4th year (75)	86	9	5

Hyde (1959) also repeated this experiment using plasticine balls and obtained similar results. Likewise Elkind (1961) found that by 8–9 years of age, three quarters of his sample had a satisfactory grasp of the conservation of mass. Furthermore, Price-Williams (1961) reported his findings among illiterate bush West African children of the Tiv tribe who were tested on the question of the conservation of both continuous and discontinuous quantities. Earth and nuts were used as examples of continuous and discontinuous quantities respectively. His results show that the progression of the idea of conservation followed that found in European and other Western children, although there is some doubt about the actual age at which conservation is reached on account of the difficulty of obtaining the ages of the children. Thus the stages proposed by Piaget have been confirmed, and the answers given by the children often agree closely with those reported by him. On the other hand the work shows that the development of the concept is much more complex than Piaget reckons. There is no doubt then that up to about 7–8 years of age many children will believe that if a substance changes its shape it is likely to change in amount.

E

In a further experiment Lovell and Ogilvie used a rubber band. The children were asked if there was the same amount of rubber in the band when it was stretched, as there had been a moment earlier, when it was unstretched. It was found that about one-third of those who were non-conservers in the experiment involving plasticine were conservers in the experiment employing the rubber band. Hyde also found that some children who were non-conservers in the test using plasticine balls were conservers when a liquid was poured from one vessel into another of different shape. Again, Hyde found that when liquids in vessels A and B, which certain children had agreed were equal in amount, were poured into glasses C and D respectively, they agreed that the amounts in the latter vessels were equal, in spite of the great perceptual differences (Figure 8).

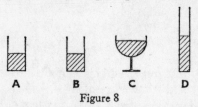

Figure 8

Yet, when the liquids in A and B were matched again for equality, and the liquid in B was poured into a number of smaller vessels (Figure 9) which were similar to one another, some of

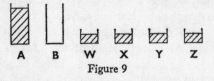

Figure 9

those who had previously agreed to conservation, now denied that the total amount of liquid in W, X, Y and Z was equal to the amount in A. Beard (1957) also found, among the 60 6–7-year-olds she tested, that some who were non-conservers when comparing balls of plasticine could conserve when water was poured from one vessel to a number of smaller ones.

The evidence from these experiments strongly suggests that as far as continuous quantities are concerned, children who are conservers of substance in one situation are not inevitably conservers in another. On this point the more recent findings seem to lead to conclusions at variance with those of Piaget. The interpretation usually given to his view is that once the concept of conservation of substance has been attained, it holds in all situations involving conservation of substance, whether continuous or discontinuous quantities are used. If Piaget is understood correctly, the evidence supplied by Lovell and Ogilvie, as well as that of Hyde and Beard, does not support his view, which in our opinion, hides the distinctive traits of child thinking. It seems, rather, that the concept is applicable only in highly specific situations at first, and that it increases in depth and complexity with experience and maturation. Piaget's view that the child arrives at the concept because he is able to argue logically may or may not be correct. It is believed that he underestimates the part played by the experience of the child in experimenting with water, sand, plasticine, etc, in many and varied situations. Sheer experience of the physical world seems to be affecting conservation more than Piaget reckons, and this theme will be developed further, later in the book. It is equally likely that the concept of conservation of substance— indeed, any concept—grows out of the interlocking of several organizations of past impressions that normally remain outside consciousness (schemata), which in turn grow out of many and varied experiences. The child may then invoke logical argument to justify the attainment of a concept which was, in fact, attained on grounds other than logical. Or, it is equally likely that experience and the development of logical thought abet one another and together give certainty.

It is not possible to say how far the inability to conserve substance affects the lives of children and the less intelligent adolescents and adults. Most people, except for those suffering from moderate or severe mental subnormality, probably become aware that the amount of water in a bucket or of cement in a bag remains the same if nothing is added or taken from it or if it is not disturbed in any way. But some of the least able, or least wary, may continue to be uncertain of the concept of invariance of substance even in very simple situations and the uncertainly may influence their lives in ways we know not. For example, Lovell and Ogilvie report that some children will pay more for a piece of toffee when it is in one shape than when it is in another. Such persons may be more often deceived by their perceptions than the majority are, and they are probably more susceptible to deception practised by the unscrupulous.

REFERENCES

BEARD, R (1957). 'An Investigation of Concept Formation among Infant School Children'. Unpublished PhD Thesis: University of London Library.

ELKIND, D (1961). 'Children's Discovery of the Conservation of Mass, Weight, and Volume'. *J Genet Psychol*, **98**, 219–227.

HYDE, D M (1959). 'An Investigation of Piaget's Theories of the Development of the Concept of Number'. Unpublished PhD Thesis: University of London Library.

LOVELL, K and OGILVIE, E (1960). 'A Study of the Conservation of Substance in the Junior School Child'. *Brit J Educ Psychol*, **30**, 109–118.

PIAGET, J (1950). *The Psychology of Intelligence*. London: Routledge and Kegan Paul.

PIAGET, J (1953a). *The Origin of Intelligence in the Child*. London: Routledge and Kegan Paul.

PIAGET, J (1953b). *Logic and Psychology*. University of Manchester Press.

PIAGET, J (1955). *The Child's Construction of Reality*. London: Routledge and Kegan Paul.

PRICE-WILLIAMS, D R (1961). 'A Study Concerning Concepts of Conservation of Quantities among Primitive Children'. *Acta Psychologica*, **18**, 297–305.

CHAPTER SIX

The Concept of Weight

It is useful to begin the study of the concept of weight by defining the term 'weight' itself. The Shorter Oxford Dictionary defines it as 'The quantity of a portion of matter as measured by the amount of its downward force due to gravity'. In other words 'weight' is a result of the gravitational pull. It is not the same as the amount of matter; the amount of matter in a body remains the same whether at 10,000 feet above mean sea level or 10,000 feet below; but its weight changes. The body will weigh heavier at the bottom of a coal mine than on the mountain top, if weighed on a spring balance, although the quantity of matter will naturally be the same. This distinction between amount or quantity of substance, and weight, will be made again later on; for, according to Piaget, the concept of conservation of weight develops a couple of years later than the concept of conservation of substance.

How then does the child acquire this concept of weight? In his pre-school days he will hear words like 'heavy' and 'light' being used, but it is not until he has picked up objects and by means of his muscle sense felt this gravitational pull—he does not of course know it as such—that he can have any idea of the meanings of words associated with weight. Let it be repeated, the concept of weight commences to develop through muscle sense, and the 'picking up' of objects or 'carrying loads on the back' comes before the use of scales. The child, then, needs much experience of comparing weights, using his own muscles. These then are some of the activities that will help:

1 Children pick up objects and learn to distinguish by

muscle sense which is the heavier. This game can be
played with the eyes closed.

2 Children examine groups of objects, each group consisting
of two objects. They then say, or write down, which of
the objects in each group is the heavier, and check their
answer by lifting the objects.

3 Miming. Children can mime a scene in which one of
them carries a heavy sack on his back and another a light
one. Another can blow a feather into the air; and
another struggle with a heavy bucket of sand.

When adequate experience has been obtained of weight
through muscle sense we may proceed to the use of scales;
indeed we must do so. With scales the child can find which is
the heavier of two objects without having to handle them.
At first children do not use standard weights, they balance a
stone in one pan against, say, sand in the other. This experiment should be repeated with sand replaced by shells, lead shot,
nuts, etc. From such activities the children learn that a small
amount of one substance weighs as much as a far larger amount
of the other, and from this point one can lead on to the need
for standard weights.

The 1 lb weight is the best one to begin with, since it is very
widely used in our society. If it is not possible to have a
number of these weights for the children to handle, make such
weights from a bag of suitable material filled with sand, or
dried peas or beans, etc and clearly marked 1 lb. The pupils
should then get plenty of practice in using the scales to weigh
out 1 lb of sand, shells, nuts, dried peas, etc. This activity
can be linked with the work of the shop, since the materials
which are suitable can be weighed, put into bags, and sold.

Soon the need will arise for a unit smaller than the lb. This
is the moment to introduce the ounce. Once again there should
be a number of 1 oz weights available, with which children weigh
out ounces of various objects. Thus they learn to weigh
materials in pounds and ounces. At this point a record of the
weighings should be made in their work books; e g weight of
parcel 1 lb 7 oz. At a later stage the $\frac{1}{2}$ lb and $\frac{1}{4}$ lb weights must
be introduced. If children have the concepts of weight and of

'$\frac{1}{2}$' and '$\frac{1}{4}$', they should appreciate that these weights are respectively the equivalent of 8 and 4 oz. It is also possible to make up bags containing 14 and 28 lb of sand to give children experience of lifting a stone and a quarter. Hundredweights and tons, however, can only be discussed in terms of what adults and lorries respectively can carry.

Children should also be introduced to the spring balance. By attaching various weights at the bottom of a thin piece of rubber, a suitable piece of elastic, or a spiral spring, and noting the elongation produced, one can demonstrate the principle of the spring balance. Furthermore, these experiments neatly illustrate weight as being the property of a body due to the pull of the earth.

PIAGET'S VIEWS ON THE GROWTH OF THE CONCEPT OF CONSERVATION OF WEIGHT

We now turn to discuss what is known about the development of the concept of weight in children in greater detail. In a typical experiment carried out by Piaget and his students a ball of clay was taken and either elongated to a 'sausage' or cut up into sections. The child was then asked whether the 'sausage', or the sections taken together, weighed as much as the original ball. Once again Piaget claims that children go through three stages in which (1) they deny conservation of weight; (2) admit to it sometimes (transition stage); and (3) readily agree to it and stand firm in their conviction (conservation stage). In the first stage, the child may reply—'It is lighter, because it is thin now', or 'It is heavier, because there are more pieces'. It seems as if the child centres on only one aspect of the transformation and that his thinking is still too influenced by perception; there has been no intellectualization of the problem and because of this failure there can be no real thought about it. He has not grasped, in Piaget's view, that because the 'sausage' can be returned to a ball and the pieces of the ball can be put together again, there must be conservation of weight.

In another experiment, there were four equal rectangular bars of tin A, B, C, D, coloured differently, and a piece of lead L. All five objects were of the same weight, but the amount of

BASIC MATHEMATICAL AND SCIENTIFIC CONCEPTS

substance in L was very different from that in A, B, C, D.
After an actual experiment using a balance the child was asked
(in suitable language):

1 If A balances B, and B balances C, how do A and C
   compare for weight?
2 If A balances B, and B balances L, how do A and L
   compare for weight?
3 If A balances B, and C balances D, what is the relation
   between the weight of A + C, and B + D?
4 If A balances B, and C balances L, what is the relationship
   between the weight of A + C, and B + L?

At the first stage, which corresponds to the stage of non-
conservation of weight, no correct conclusions were arrived at
by any of the children. At stage two, corresponding to the
stage of transition, problems were solved involving the similar
bars A, B, C, D, although the children failed all problems
that brought in the lump of lead L. This happened although
the age of the children suggested that the concept of quantity
had been formed. Perhaps, at this stage, children may solve
the problem of similar bars by simply considering the amount
of matter. If so they executed the logical operations correctly
in the case of substance (where conservation or invariance was
understood), and obtained the right answer by inadequate
reasoning. Hence for Piaget, the intellectualization of the
problem in relation to weight had not gone far enough at this
stage. They had no concept of invariant weight, and therefore
could not solve problems involving L. The child still remained
too much dependent on subjective experience. He could be
shown in all apparent clarity that the lead bar balanced the blue
tin bar, and that the blue tin bar balanced the red tin bar,
but he could not derive that the lead balanced the red tin bar.
At stage three, where the child was able to execute logical
operations in relation to weight, because of added experience
and maturation, he could arrive at the concept of invariance of
weight. At this stage, too, he could perform the operation of
transitivity, viz he could derive that if A = B in weight, and
B = L in weight, then A = L in weight. Stage three does not, in
Piaget's view, come until about 9–10 years of age (Piaget 1941).

72

CONCEPT OF WEIGHT

FURTHER EXPERIMENTAL EVIDENCE ON THE DEVELOPMENT OF THE
CHILD'S CONCEPT OF CONSERVATION OF WEIGHT

In experiments carried out by one of the writer's students,
(Lovell and Ogilvie 1961), a large number of junior school
children were shown two balls of plasticine of diameter 1½ and
2 inches respectively, and they had to judge which was the
heavier using their hands or scales as they wished. Actually
the smaller ball, which had some lead at its centre, was clearly
heavier than the larger. These balls will later be referred to
as R1 (the heavier but smaller ball) and R2 respectively. The
latter ball was then rolled out to form a 'sausage' and the
children were questioned at length, and individually, about
the weight of the 'sausage' and the ball. Table 2 shows
the percentage of children at each year of the junior school who
were found at the conservation, transition, and non-conservation
stages.

TABLE 2

Year	Stage 3 Conservation %	Stage 2 Transition %	Stage 1 Non-Conservation %	No of Children Tested
1st year	4	5	91	57
2nd year	36	36	29	73
3rd year	48	20	32	66
4th year	74	13	13	168
				364

The stages proposed by Piaget are again confirmed, and the
answers sometimes agree with those reported by him. Moreover, a comparison of Table 2 above and Table 1 shows that
the concept of conservation of weight usually develops later
than conservation of substance, although this may not be so
in individual cases (see also Hyde *op cit*).

But the experimentation went far beyond this point, and a
number of findings were clearly established that apparently
have been overlooked by the Geneva school. These may be
briefly summarized as follows:

1 Some 48 per cent of the children at the transition stage,

73

and some 46 per cent at the non-conservation stage, could show reversibility of thought, for they gave evidence of being aware of the weight relationship between R1 and R2 at the beginning of the experiment. It follows that reversibility of thought itself is not a sufficient condition for conservation, although it may be a necessary condition. This clearly parallels the results of Lovell and Ogilvie (1960 Table 5), who found that 75 per cent of those at the transition stage, and 50 per cent of those at the non-conservation stage of substance, showed evidence of reversibility.

2 The children were then shown a third ball G which was heavier than R1. They were asked individually, and in suitable language, what the relationship was between the weights of G and R2, if G was heavier than R1 and R1 was heavier than R2. Contrary to what one might expect from Piaget, 67 per cent of those at the transition stage and 53 per cent of those at the non-conservation stage could perform the operation of transitivity.

3 A number of questions of the type given below were put to 114 of the conservers—so designated on Piaget's criterion. 'What happens to the weight of this ball (R2) if I squeeze it down to the size of this one (R1)'.

'What will happen to the weight of this ball (R2) if I leave it in my cupboard for a time?'

'If the plasticine gets harder does it get heavier?' 'Or does it get lighter?'

The following numbers of conservers said, in effect, that *harder* indicated *heavier*, although this was not believed to be true by all subjects in relation to all substances: 1st year 2; 2nd year 16; 3rd year 18; 4th year 42; making 78 in all.

4 Children who were conservers when rolling plasticine were asked: (*a*) What would happen to the weight of a piece of butter if it hardened? (*b*) If some water was cooled and it changed into ice, would the ice weigh more, the same, or less than the water from which it was formed? (*c*) What would happen to a lump of clay if it got harder? From answers to these questions, and from the varying

percentage of conservers obtained in each case (much the highest in (*a*)), it is perfectly clear that there is no automatic transfer of reasoning from one substance to another. In the case of butter there is much practical experience of it, through shopping, in which weight is involved; the child also notices it getting harder and softer in the house with changes in temperature. Whereas the child may manipulate plasticine more than butter, the weight as such is rarely considered.

The conclusion to be drawn is that few children in the junior school have generalized their ideas about weight; that is, few of them have conceived of it as a factor abstracted from subjective feelings and linked with quantity and only quantity—so that they will retain conservation of weight provided the quantity of substance does not change. Until the child has learnt from experience that warming, cooling, squeezing, hardening, ageing, lengthening and so on do not alter the weight, he will not conserve weight in the widest sense. In spite of the fact that the child can give evidence of reversibility in relation to weight, that is, he can state the relationships between the weight of balls R1 and R2 at the beginning; and in spite of the fact that he can perform the operation of transitivity, he will not necessarily conserve in the general sense, since he may *believe* that some process one of the balls undergoes actually changes its weight. Piaget's tests of rolling a ball into a 'sausage' or of cutting a ball into slices by no means told the whole story. The tests simply showed up those who conserved in a particular kind of experiment using a particular medium. With change of medium and circumstance the percentage of children at the various stages who conserve weight may well change. The kind of experiments described here might be undertaken in the junior school to help the child to learn that weight depends solely upon the amount of material, at least for all practical purposes.

Summing up then, we may say:

1 The concept of the invariance of weight develops in part through prolonged and varied experience of the physical world.

2   Piaget's are only one kind of experiment that could be used to test the development of the concept.

3   The type of experiment used by Piaget, and other types of experiments described here, can be in themselves learning situations.

The concept of conservation of substance arises earlier than that of weight since quantity is under immediate visual perception, whereas weight is not.

As far as the acquisition of transitivity of weight in 5–7 year-old children is concerned, Smedslund (1963a) has concluded that the task of ordering three objects in a series by means of a balance seems to be the most effective procedure in bringing about cognitive conflict and successful cognitive reorganization.

In a further study Smedslund (1963b) studied the acquisition of the transitivity of weight in pupils aged 7 : 6 to 9 : 3.   He attempted to test two views:

(i)   Transitivity results from repeated observation that if A weighs more than B, and B weighs more than C, then A weighs more than C (learning theory).   It is important to have many sequences of the type $A>B - B>C - A>C$, and observation of $A>C$ are necessary on this view.

(ii)   Transitivity results from the internal reorganization of schemas.   The reorganization is presumably initiated by repeated uncertainties or problems.   He also studied the effects of free practice in weighing compared with fixed practice, i e objects had to be weighed in a fixed order.

Smedslund thinks that the second view is the correct one, but frequency of problem or cognitive conflict was crucial. Also, while the younger children seemed to benefit a little more by free practice, the older children profited equally from free and fixed practice.

In his later monograph (Smedslund, 1964) he stresses that other things being equal, conservation develops before transitivity.   His views do not run counter to our findings since in our experimental work 'other things' were not 'equal'.

It will be appreciated that a child may work exercises involving lb, oz, qr, etc, without conserving weight.   Provided he knows the appropriate tables, can manipulate the four rules of number, and can comprehend the question, he can work the exercise, since conservation used in the widest sense is not involved. But in problems of science which involve changes of state, density, etc, his thinking will be based on his pre-conceptions and may lead him into gross errors.

In connection with this chapter readers might study an article by King (1961). His data were not obtained by individual questioning but by means of a questionnaire.

REFERENCES

KING, W H (1961). 'The Developments of Scientific Concepts in Children'. *Brit J Educ Psychol*, **31**, 1–20.

LOVELL, K and OGILVIE, E (1961). 'A Study of the Conservation of Weight in the Junior School Child'. *Brit J Educ Psychol*, **31**, 138–144.

PIAGET, J and INHELDER, B (1941). *Le développement des quantités chez l'enfant*. Neuchâtel: Delachaux and Niestlé.

SMEDSLUND, J (1963a). 'The Acquisition of Weight in Five-to-Seven Year-Old Children'. *J Genet Psychol*, **102**, 245–255.

SMEDSLUND, J (1963b). 'Patterns of Experience and the Acquisition of Concrete Transitivity of Weight in Eight-year-old Children'. *Scand J Physiol*, 4, 251–256.

SMEDSLUND, J (1964). 'Concrete Reasoning: A Study of Intellectual Development'. *Monogr Soc Res in Child Develpm*, No 93, Vol 29.

CHAPTER SEVEN

The Concept of Time

AUGUSTINE in his *Confessions* had this to say about time: *
'For what is time? Who can readily and briefly explain this?
Who can in thought comprehend it, so as to utter a word about
it?' The Bishop of Hippo had long reflected on the nature of
the concept, as had earlier Greek philosophers, and decided
that it is not an easy one for the human mind to understand.
In this century many people, including Burt, have pointed out
the difficulty the child has in making a 'time synthesis'.
Indeed, it seems that the concepts of both space and time are
only slowly built up and involve the construction and elaboration
of certain essential relationships. Today it is more important
than ever that we think about the concept and help our pupils,
as far as we can, to grasp its significance, since *time* is one of the
fundamental concepts involved in mathematics and science.
As we have already said, all our concepts, even at the adult stage,
get wider and deeper with added experience as long as the brain
and mind maintain their integrity, providing prejudice or
emotion does not have a narrowing effect on categorization.
Thus, in general, the 18-year-old is expected to have a better
understanding of time than the 8-year-old, although the latter
should have some grasp of the concept.

Children of 3 or 4 years of age have a sense of timing but
this is not a concept of time. A young child runs, extends his
arms and tries to catch a ball. Again, a dog will watch a
bouncing ball and catch it in his mouth, and the bird of prey
will show a magnificent sense of timing as it swoops on its

* Augustine. *Confessions*, Book xi, 14.

quarry. But though some animals, as well as children have a sense of timing, there is no evidence that, either in them or in the young child, there arises, in consciousness, any understanding of time.

TIME PERCEPTION

It was earlier pointed out that percepts lead, in ways we little understand, to concepts. That is as true of the concept of time as of any other concept. The common events of everyday life become integrated into perceptual patterns in some way. Indeed, E G Boring, a distinguished American psychologist, has suggested that time perception has five bases. In his view the child:

1 gets some perception of succession, or how stimuli follow one another, as, for example, when he runs a pencil over the teeth of a comb at different rates;

2 acquires some perception of continuity, as when he notes some action continuing until it stops; e g a slowly turning wheel;

3 obtains some idea of temporal length from the differing perceptions involved in, say, the playing of a long and short musical note;

4 learns to respond to the environmental signals of the 'present', such as the feelings of hunger associated with an empty stomach;

5 acquires the ability to perceive patterns of successive stimuli. This ability to feel rhythm may have a physiological basis.

DEFINITION OF TIME

Typical dictionary definitions of the word *time* read as follows: 'a limited stretch of continued existence'; 'the interval between two events'; 'the interval through which an action, condition, or state continues'. From these we may fairly say that before a child can have some grasp of the concept of time, the order of succession of given moments on the time continuum has to be co-ordinated with the period that extends between these moments. For example, when a child is watching a simple pendulum swinging, he does not have much concept of time until the points on the time continuum (i e the instants when the pendulum is at the extremes of the swing) are co-ordinated mentally with the period in the time continuum (the period of the swing of the pendulum from one position to another). The child is not, of course, conscious of this co-ordination as such, but he is aware of two instants and of the continuous nature of the interval between them.

TELLING THE TIME

That a child can tell the time on the clock does not necessarily imply that he has a concept of time. Telling the time is no more and no less than dial reading. One may read a dial without having any concept of what it is that the dial registers. This point is easily illustrated at adult level. The instrument panels of a modern air liner contain many dials. Only a fraction of the number of people who could be taught to read all the dials, could be brought to the position where they grasped the concepts involved in what the dials measure. Now when the child can read off the time on the clock, it is possible that he has had enough experience, and has sufficient neurological maturation, to have some concept of time. Furthermore, learning to tell the time may help the child to have some understanding of time. It must not be thought, of course, that helping children to tell the time is an unimportant task. On the contrary, it is the teacher's clear duty to help all children to do so, since the ability to tell the time is an almost indispensable social asset in our culture.

PRIMITIVE AND EARLY CHILDHOOD CONCEPTS OF TIME

It is instructive to note that in primitive societies the words used to express time reflect no more than the main events in the individual's day. When working with cattle, say, the main events of the day might divide it into time divisions such as *watering time, coming-home time, milking time.* Again, the divisions of the year are arranged in terms of important events, e g *planting time, harvest time,* etc. Primitive man uses, as it were, moments of time embedded in a continuum of action.

This is different from our idea of time. Our concept of time has continuity and consistency, which are features of a completely abstract time that can be quantitatively measured.

In the case of young children too, time is marked off by isolated and distinct events and actions, most of which arouse strong feelings. At first time is intermingled with impressions of duration which are inherent in attitudes of expectation, effort, desire, success, failure and satisfaction. According to Werner (1957) when a 2-year-old child uses a word such as 'bath' he indicates a wish in which the time element of the near future is implied. When he uses the words 'all done', satisfaction is expressed about some activity just completed and past. Time is embedded in a series of events—in a continuum in which space and time are not differentiated.

However, most children move away from this very egocentric apprehension of time, and the duration of the events in which they are very personally involved is put into closer and closer relationship with the events in the external world. Time then becomes universal and 'flowing', the same for everyone; it is a time that is continuous and contains the general and objective event, and not just the events that are of immediate importance to the child. The sequence of a child's life is then inserted, as a lived sequence, into a whole series of events which constitute the history of his environment. But progress is slow and there are pitfalls. Indeed the slowest learners never grasp a developed concept of time.

SOME EXPERIMENTS INVOLVING TIME CONCEPTS

It must be realized, then, that a child can use *time words* to describe concrete things or events without having much concept of time. Ames (1946) tried to establish the growth sequence of the development of the sense of time by analyzing the spontaneous verbalizations and answers to a series of questions, such as, 'What day is it today', 'What time is it now?' Using small groups of children from 2 to 8 years of age, Ames found that the understanding of time words and their use came in a fairly uniform sequence. The sequence was roughly the same as for all children, although the age at which a particular child could understand or use a specific term varied. For example, she says that children (in the U S A) can use the terms *morning* and *afternoon* by 4 years of age; they know what day it is by about 5 years, and can tell the time by about 7 years of age. Furthermore, it was shown that the child can use terms like *winter time* and *lunch time* before the word *time* itself.

Such studies suggest that a vocabulary of time words is built up by a process of association. When the child plays or it is light, it is *day*, when he goes to bed or it is dark it is *night*. When he gets up it is *morning* or *breakfast time*, when it is cold or snowing it is *winter*, when it is hot it is *summer*, and when the corn is being cut it is *autumn*. It is, of course, essential that we do all we can to help the child to build up an active vocabulary of time words; and it is certain that an understanding of relevant associations and use of such words helps the child, to some extent, to develop his concept of time. But these associations, and the use of the words, do not ensure that he has a grasp of the concept. Naturally, when he comes to use a word like *morning* with some awareness of the interval between breakfast and lunch, he is beginning to have some understanding of time. It is significant that in Ames's study it was found that time words could be used before the word *time* itself. When the child can use this word it is likely that he has some concept of time, although even here we must be careful. Without understanding the word *time* as such he might say 'What time is it?' or 'It is time I went home', and be merely repeating expressions he has heard adults use.

Another relevant experiment at this juncture is that reported by Sturt (1925). She (with her collaborator E C Oakden) asked children in one part of their study these two questions:

1   What day of the week is it?
2   What day is it at X (a nearby town) now?

The percentage of children giving the correct answer is shown below:

	Age in Years			
	4	6	8	10
Question 1 . .	27	87	97	100
Question 2 . .	14	60	64	86

Most children over 6 years of age realized that it was the same day of the week all over the same town, but even at 10 years of age the fact that the day is the same in all English towns was not realized by 14 per cent of this sample.

The study of Bradley (1948) suggests that the knowledge of time is acquired in this general sequence:

1   time related to personal experience; e g time in relation to child's own age, morning and afternoon(6 years of age);
2   conventional time words used in the calendar and the organization of the week (Well understood by 8 years of age). There is a tendency for growth to occur outwards, viz week, month, year;
3   time involving extension in space and duration; e g time elsewhere, time since holidays, how long since child started school. The extension of an understanding of time in space agrees with that of Sturt (*op cit*).

Springer's (1952) study is of help because it suggests the stages into which the instruction may be divided when helping children to tell the time by the clock. She studied 89 children aged 4–6 years of age who were attending a kindergarten, and who had received no school instruction in the tasks which she set them, although they had had incidental experiences and informal instruction in the home. They were asked, in individual interviews, to tell the time of certain activities in their daily lives, e g breakfast time; to tell the time by a clock, to set a clock, and to answer questions about the hands of a clock. Although there was overlapping the following general sequence of development emerged:

1   the child learns the time of activities in his daily life. But even this goes through a sequence. To such a question as 'What time do you have lunch?', he replies with a descriptive term like 'Afternoon'. Or he describes a sequence of events, e g 'After lunch I have a sleep and go home.' Next an unreasonable time is given; then a reasonable but somewhat inaccurate time; and finally accurate time;
2   he can tell the time by the clock first by whole hours then by halves, then by quarters;

83

3 he can set the hands of a clock to whole hours, half hours, quarter hours;
4 he can explain why the clock has two hands.

Springer's groups were drawn from above average socio-economic groups; in children from poorer backgrounds the sequence may be delayed.

PIAGET'S EXPERIMENTS INVOLVING THE CONCEPT OF TIME

The concept of time is not an easy one, as there are few specific clues. Piaget's experiments show how the child has difficulty in grasping the meaning of time. In one experiment two dolls raced on a table in front of the child. They started at the same time and a 'click' was made at the instant of their starting. One doll moved faster than the other. They stopped simultaneously with a 'click', one doll being ahead of the other. Piaget maintains that up to 6 years of age the child denies that the dolls stop at the same time, or that the length of the time for which each was running is the same, in spite of the fact that he can admit that the dolls started at the same time. The child said that the faster moving doll took longer and stopped later. Between 6 and 7 years of age children admitted that the dolls started and stopped at the same time, but for them the faster moving doll still took longer. It was not until 7–8 years of age that children admitted that the dolls started and stopped together, and that the period of movement was the same for each doll. At this stage there is co-ordination in the child's mind between points and period on the time continuum.

In another experiment a container of water had an outlet with a control valve. After leaving the container the tube divided into two branches forming an upside-down Y. The same amount of water flowed through both branches and because of the valve, the flow of water in one tube both started and finished at the same time as that in the other. Two empty glasses of different width were placed under the two outlets, and the water flowed until the narrower glass was full, the wider glass remaining only partly full. At 4–5 years of age, Piaget maintains, the child thinks that the water in the tube which completely filled the narrower glass flowed for a longer time than the water in the tube which partly filled the wider glass. In this as in other examples at the intuitive stage of thinking, thought seems based on perception in which one aspect of the situation is centred on at the expense of other aspects or dimensions.

These and other experiments led Piaget to suggest that at first the child's ideas of time are all mixed up with his ideas of space and spatial changes. It is not until time has been intellectualized (i e until instants and points in the time continuum are co-ordinated in the mind), that it becomes an invariant quantity independent of quickness of movement, distance moved through,* and position. Indeed, Piaget contends that the time concept depends upon the child being able to build up coherent systems of logical thought, such as those that lead to the conservation of quantities. For Piaget the concept of time will begin to emerge at the same time as other concepts of the physical world.

FURTHER EXPERIMENTS RELATING TO THE DEVELOPMENT OF THE CONCEPT OF TIME

At the University of Leeds Institute of Education a number of experiments have been carried out, under the writer's direction (Lovell and Slater, 1960), similar to those suggested by Piaget. Seven experiments were used, some of which will be described here. They were undertaken, individually, by:

(a) ten average to bright children in each of the age groups 5–9 inclusive in a primary school, making 50 pupils in all;

(b) an equal number of pupils in junior and secondary E S N special schools whose ages ranged from 8 to 15 years.

For present purposes we are not concerned with the results obtained from the latter children.

Two of the experiments involved:

1 the flow of water from an upside down Y tube into (a) a test tube, and (b) a wider but shorter beaker. The

---

* For most purposes we may ignore relativity theory, which shows that the time interval must be measured for each observer for himself. Space and time cannot be regarded as separate, each is part of the whole and dependent on the other.

apparatus was much the same as that used by Piaget and already described,

2 two dolls racing across a table in much the same way as in Piaget's experiments. A red doll started from A and finished at B, while a yellow doll started from C and finished at D on a parallel course. Both started and stopped, with a 'click', at the same instant.

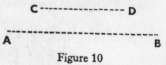

Figure 10

In both experiments the children were asked, in a suitable manner, questions such as:

1 Did the water start flowing into both glasses at the same time?

- Did the water stop flowing into both glasses at the same time?

Did the water flow into each glass for the same time?

2 Did the dolls start walking at the same time?

Did the dolls stop walking at the same time?

Did one doll start walking later than the other?

Did both dolls walk the same distance?

The results obtained from the two experiments described above, and from others, revealed:

(a) greater accuracy in the perception of simultaneity, and a greater understanding of equality of synchronous intervals and of order of events with increasing age. To a number of questions there were 100 per cent correct replies at 9 years of age;

(b) these understandings came earlier in the case of the 'water flowing' than in the case of the 'doll racing' experiment. Furthermore, the idea of equality of synchronous intervals came earlier in another experiment involving a siphon-than in either of the two experiments described in more detail. Therefore, a specific concept does not develop in all situations and in all media at the same age (for any one child); contrary, it seems, to Piaget's belief;

(c) in our population of primary school children the concept of time developed later than Piaget suggests; this in spite of the fact that our infant and junior school population contained no backward children;

(d) many of the younger children thought that the water flowed for a longer time in the test tube which filled than in the beaker which only partly filled. Many of them also maintained—some even of the 9-year-olds did— that the red doll, which travelled faster and further, walked for a longer period of time than the yellow doll.

From the above it is clear that the concept of time, considered in the scientific sense, in which order of events and equality of synchronous intervals are concerned, grows slowly, and is not available to the child equally well in all situations at once. Certainly, there are good reasons for believing that, as Piaget suggests, time and space are undifferentiated at first. In the case of the E S N special school children, the results clearly indicated that their concept of time is still developing between 12 and 16–17 years of age, as compared with 5–10 years of age in children outside such schools, except that in the perception of simultaneity the performance of the senior E S N special school children was comparable with that of the primary school pupils.

Lovell and Slater also confirmed Piaget's findings that for 5-year-olds, age is dependent upon size. Thus if tree A is planted later than tree B, but grows more quickly so that it becomes taller than B, then tree A is judged to be the older.

A useful review of many aspects of the psychology of time has been provided by Fraisse (1964).

Children may be quite capable of working the four rules of number in relation to days, hours, minutes and seconds, without having a concept of time in which instants and intervals on the time continuum are co-ordinated in the mind. They might equally well have been taught to work the four rules involving the table 60 sics =1 mub, 60 mubs = 1 huz, 24 huzes =1 dir.

ESTIMATING TIME INTERVALS

When the child has begun to have some understanding of time, his actual estimates may be poor. Even when the concept

87

is better developed the accuracy of his estimate may vary with mood or interest, so that 20 minutes in which he is very active or happily engaged in some creative work may be estimated as a much shorter period than one of 20 minutes during which he is bored. The same is true of adults. It seems that periods of time filled by many different kinds of thoughts appear shorter, those filled by few different kinds of thoughts appear longer. When we are very interested we may have many thoughts but they are centred around the interest, whereas when bored our minds wander in reality or in fantasy and the time period appears longer. Varying estimates of the same period provide good opportunities to bring home to the pupils the need for fixed units of time, e g second and minute.

Children should be given opportunities to practice estimating the time they spend in specific activities, say, in reading, running playing, singing, listening. They should check their estimates against the second or minute hand of a watch. As a further means of helping children to estimate a second and minute it is useful to have a simple pendulum in the classroom. All that is needed is a piece of string 39 inches long attached to a nail in a wall, and at the loose end a small piece of metal to act as a bob. The time taken to move from one extreme of the swing to the other extreme position is approximately a second $\left(T=2\pi \sqrt{\dfrac{l}{g}} \right)$, with 2 seconds elapsing before the bob returns to its former position. It will be much later before the child has an understanding of the period of 1 hour. At first, it will be vaguely grasped, perhaps as the time interval between playtime and the end of morning school. Furthermore, many studies since the turn of this century have shown that average children have little understanding of the time period of 1 year, until they are at least 9 years of age (cf Bradley).

Goldstone (1958) asked 190 children aged 6–14 years to count 30 seconds aloud, and to themselves, at the rate of one count per second. Their concept of one second was then estimated from this count. The 8 year, and older children, together with young adults, made good estimates of one second when counting to themselves, although the 6- and 7-year-olds and older adults made shorter estimates. Counting aloud, which involved muscle activity, resulted in longer estimates. It is possible, then, that kinaesthetic cues may play a part in learning and appreciating units of duration.

CONCLUSION

We cannot be sure how much we can help children to develop their concept of time, nor do we know the means most likely to help them. However, it is probable that watching or listening to any activity that has a clear beginning and ending is likely to be a learning situation, as far as the time concept is concerned; e g watching the sand in an 'egg timer', or listening to a musical note being played.

It is certain that children can be helped to build up a vocabulary of time words, and to tell the time. Moreover, once some concept of time has been developed, children can be given practice in estimating time intervals. Our teaching is bedevilled by the fact that children can use words which often lead us to an exaggerated notion of their understanding. While it is the clear duty of teachers to help children to build up a vocabulary of time words which they can use meaningfully, and to help them to tell the time, these attainments, in themselves, do not ensure that children have any grasp of the meaning of time itself. It seems that the very rhythm of life helps the child to develop his concept of time. Darkness follows light, and the child's mealtimes occur, in fairly regular sequence; he goes to school each morning after breakfast, plays at certain times and so forth. These and similar everyday routines may facilitate the development of that awareness whereby the child links in his mind the successions of events with the interval that separates them.

It might be objected that the adequate use of time words and competence in telling the time are all that is needed for most people, and all that many people possess. The reply to that is that, though we are indeed grateful if the slowest learners acquire these skills, the majority of our pupils need much more. For them, time is an important concept. Unless they have some understanding of the concept, the terms *second*, *minute*, *hour*, etc can have very little meaning. Furthermore, when they are older, concepts such as velocity, acceleration, per unit time, distance-time graph, velocity-time graph—all of which frequently occur in mathematical and scientific education today —are quite beyond their understanding.

CONCEPT OF VELOCITY *

It will be useful to discuss briefly the development of the concept of velocity immediately after our treatment of the concept of time. Here we have to rely very heavily on the work of Piaget.

We may begin by suggesting that the first glimmer of under-standing of the meaning of the word *faster* is by association. From his play, and from what he hears adults say in real life situations, the child gathers that if one person or object in movement overtakes another person or object, the former is said to be going *faster*. He does not attach the same significance to the word that the educated adult does; namely that of a distance-time relationship. If one child runs past another in play, both of course going in the same direction, or if one motor car overtakes another, the former in each case is said to be going faster. In the child's mind it is merely the word to denote the movement of one of the objects when there is a change of relative position. Likewise he learns the word *slower* as being the word to denote the movement of the other object. This is another instance showing how a child can use a word in a correct context and yet deceive us as to his actual degree of understanding. The same is most likely to be true of the less intelligent adult.

Piaget has told us that Einstein, in 1928, asked him if a child's first concepts of velocity included an understanding of it as a function of distance and time, or if his first notions of it were more intuitive and primitive. Piaget and his students thereafter devised a number of experiments which have thrown some light on this question. It seems that the child is at first under the influence of his perceptions, and that it is only when he reaches the stage of concrete operational thought that he can think of velocity in terms of distance and time. A few of the experiments used by Piaget (1946) will be described to illustrate this.

---

* Velocity and speed are both taken to mean rate of movement, and the problem of vector and scalar quantities is ignored for present purposes.

1 A and B are two cardboard tunnels, one of which is obviously much longer than the other. Two dolls, moving on rods, enter the tunnels at P and at Q the same moment, and emerge at X and Y simultaneously. Although the child at 5 years of age readily admits that tunnel A is longer than tunnel B, he maintains that the dolls moved at the same speed. On being questioned he maintains that he is correct because the dolls arrived at X and Y at the same time (Figure 11).

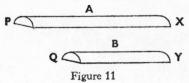

Figure 11

But if the tunnels are taken away, and the dolls are run along the floor over the same distance in full view of the child, so that they pass P and Q at the same time and pass X and Y simultaneously as before, the child admits that one doll is going faster than the other. Questioning shows that he understands that one doll caught up and, just beyond XY, passed the other, and in terms of his experience of everyday life it is going *faster*.

2 Between A and B there are two paths, one direct and one via C. If one object moves along the direct path AB, whilst a second one, moving faster, moves from A to B via C, so that both objects leave A and reach B at the same instant, the 5-year-old child judges their speeds as equal (Figure 12).

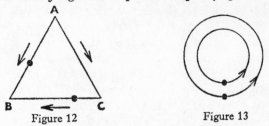

Figure 12   Figure 13

3 If two objects, moving along concentric circles keep their relative positions, young children maintain that their speeds are the same (Figure 13).

4 Two objects A and B move along parallel paths. A, the faster moving object, comes from further off, catches B, and

they both stop at that instant. The child of 5 again judges the speeds of the two objects to be the same (Figure 14).

These experiments suggest that, at first, speed is unrelated to distance covered, as far as the child is concerned. But by 7–8 years of age, Piaget maintains, the child can reassess his estimate of speed; since his manoeuvrability of thought is much greater,

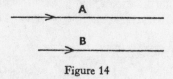

Figure 14

and he can grasp that the faster moving object in, say, experiment 4 would have gone past the slower one, had the former not stopped. It seems that the child's first notions of velocity are based on some intuitive realization that, if two objects are in motion in the same direction, the one that passes the other has the greater velocity, and the distance-time relation as such is not involved. This, it is suggested, arises in part because of the manner in which a child first gets some understanding of such words as *faster*, *slower*, *greater speed*, as has already been described. But it also seems to depend on the fact that the idea of position, or order in space, is so much easier to grasp than the concept of intervals on the time continuum. Passing is just a change of relative position, and the child at first tends to judge velocity in terms of the order of arrival at a given point. Piaget also claims that if a fast-moving object stops before it catches up one moving more slowly, the 5-year-old will often say that the object in front has the greater speed.

In a study by Lovell *et al* (1962), ten experiments of the type first described by Piaget (*op cit*) were undertaken by ten children from each of the age groups 5–10 inclusive. The pupils were selected by their teachers as a representative sample, from the point of view of ability, at each age level. Overall, Piaget's findings were confirmed, although it was found that it was not until 9 years of age that roughly three quarters of the age group understood speed as a function of distance and time rather than at 8 as proposed by Piaget. Furthermore, the notion of relative speed was in evidence in the majority of the 10-year-olds.

CONCEPT OF TIME

The experiments were also undertaken by ten children in each of the age groups 10, 12, 14 and 15 years of age in E S N special schools. Their understanding, even at 15 years of age, was scarcely equal to that of the 7–8-year-olds in the primary school when the whole range of experiments is considered. Thus even at school leaving age the majority of E S N special school pupils have a limited grasp speed in the sense of distance per unit time. This is, of course, to be expected in view of the findings of Lovell and Slater (*op cit*) in respect of time.

REFERENCES

AMES, L B (1946). 'The Development of the Sense of Time in the Young Child'. *J Genet Psychol*, **68**, 97–125.

BRADLEY, N C (1948). 'The Growth of the Knowledge of Time in Children of School Age'. *Brit J Psychol*, **38**, 67–78.

FRAISSE, P (1964). *The Psychology of Time*. London: Eyre and Spottiswoode.

GOLDSTONE, K *et al* (1958). 'Kinaesthetic Cues in the Development of Time Concepts'. *J Genet Psychol*, **93**, 185–190.

LOVELL, K, KELLETT, V L and MOORHOUSE, E (1962). 'The Growth of the Concept of Speed: a Comparative Study'. *Journal of Child Psychology and Psychiatry*, **3**, 101–110.

LOVELL, K and SLATER, A (1960). 'The Growth of the Concept of Time: a Comparative Study'. *Journal of Child Psychology and Psychiatry*, **1**, 179–190.

PIAGET, J (1946). *Le développement de la notion de temps chez l'enfant.* Paris: Presses Universitaires de France.

PIAGET, J (1946). *Les notions de mouvement et de vitesse chez l'enfant.* Paris: Presses Universitaires de France.

PIAGET, J (1955). *The Child's Construction of Reality*. London: Routledge and Kegan Paul.

SPRINGER, D (1952). 'Development in Young Children of an Understanding of Time and the Clock'. *J Genet Psychol*, **80**, 83–96.

STURT, M (1925). *The Psychology of Time*. London: Routledge.

WERNER, H (1957). *Comparative Psychology of Mental Development*. New York: Incorporated Universities Press.

CHAPTER EIGHT

Concepts of Space

AT the outset a clear distinction must be made between *perceptual space* and *representational space*. As early as 6 months of age an ordinary child can distinguish between a circle and a triangle, when they are presented visually. But it is much later that the child can represent these figures to himself in thought; that is, when he has developed some concept of these figures. Indeed, the development of these concepts emerges with maturation and experience, and, as it does so, he can increasingly communicate to others his understanding by use of symbols such as sign, drawing and writing. It seems as if this development depends on actions, as we shall see later. But as soon as he has some ability to represent to himself, in thought, spatial relationships, he can begin to execute certain actions that necessitate his taking into account spatial relations that are not directly observable. For example, Piaget showed that there comes a time when a child will go round an object, say, a screen, to search for a ball that has disappeared behind it. This behaviour implies that he can represent to himself, in thought, the spatial relationships between the ball and the screen, and the movements of his own body in relationship to these objects.

SPATIAL IDEAS IN PRIMITIVE MAN

In an earlier chapter there was a discussion of primitive man's notion of time. It may now be instructive to consider his notion of space. Werner (*op cit*) has pointed out that as early man did his everyday tasks—such as steering his canoe, and throwing a spear—his practical, everyday space, in which he

94

moved and acted, was exactly the same as the space of modern sophisticated man. But the representational space of the former —the space that was the subject of his reflective thought—was radically different from the abstract space of educated men of today.

Terms used by some primitive peoples suggest that the body itself is the source of their spatial concepts. The word *eye* might indicate *before*; the word *back* indicate *behind*; and the word *ground* indicate *under*. So their ideas of space seem to be rooted in concrete and personal situations. These situations are linked with tribal life and culture, and are often bound up with emotional life. In the case of such peoples space cannot be detached in the mind from the concrete and affective; it cannot, therefore, stand completely outside the individual, objective, measurable and abstract. There is a clear parallel here to primitive man's concept of time.

CHILDREN'S CONCEPTS OF SPACE

It seems that the child's concepts of space, like those of early man, grow out of the awareness of his own body. His first spatial knowledge of an object comes through placing it in his mouth, together with the associated tactile experience. Very slowly the space surrounding his body becomes differentiated from the body itself, and objects become known by reaching and touching them. But his space is still bounded by what he can touch. By the sixth month of life the separation, in space, of the self and the not-self, is proceeding more rapidly, and his space continues to expand. Nevertheless, space remains bound up with the self for some time longer.

Werner (*op cit*) quotes an interesting report by Scupius of a 3-year-old child which shows how concepts of space grow out of motor acts, how the child orientates and gives dimensions to space through actions. While visiting a zoo, a child was taken up by a flight of stairs and down by a different flight. The next day, on visiting the zoo again, he was taken up by the stairs he had come down previously. The child protested that these were the 'going down' stairs and the others the 'going up' stairs. Another example from Scupius involves a 7-year-old who had been taken for a long circular walk through

95

the woods by the forester. When within five minutes' walk of the house—that is, when they were almost home again—the forester showed the boy a strawberry patch. Next day the child tried to lead his mother from the house direct to the patch, but was unable to do so. Rather, he had to make the circular journey all through the woods as on the previous day, until he reached the strawberries. He relied all the time (if his verbalizations are a true guide) on a chain of associations and memories. Both these examples show that, even at 7 years of age, space can still be tied to motor acts, it can be a 'concrete space', and has not been sufficiently internalized for it to be subjected to mental operations. It remains, therefore, inflexible in the mind, not manoeuvrable.

There is a possibility that the space of the less intelligent adult in our culture is also a 'concrete' space, linked to his actions. And, indeed, the space of the educated adult is likely to revert to the concrete if he is under great stress, or if the spatial relations become too complex for him. The educated adult is likely to behave as the 7-year-old if the route through the woods is very complex.

Watts (1944) reports that the following words are learnt at the nursery level for the simplest spatial relationships: *up, down, right, left, over, under, above, below, before, behind.* He thinks they are learnt in the course of active exploration and seem to be necessary for fixing attention on what they stand for. That is to say, it is difficult for the child to discuss certain spatial relationships without these words. Watts also quotes the work of Holmes as indicating that some 80 per cent of a group of young children understood the words *on top* and *behind*; and at 5 years of age the terms *backwards* and *forwards* were understood by most of them. But not more than 60 per cent could demonstrate that they understood the difference between *tiny* and *huge*, and less than half could show that they understood the distinction between *far* and *near*.

THE CLASSIFICATION OF SPATIAL RELATIONSHIPS

1 Rigid motions. The views of Piaget and Inhelder on the development of the child's conception of space will be discussed later. Before this can be done it is necessary to

discuss how spatial relationships are classified in mathematics. Most people in our culture have had some acquaintance with Euclidean geometry in their schooldays. This geometry deals with relationships that concern *magnitudes*, such as length, size of angles, areas and volumes. In this geometry two figures are said to be congruent if they are identical in shape and size. One figure can be obtained from another by a *rigid motion* in space, in which there is only a change in position but no change in measurement. For example, triangles ABC and DEF are congruent (Figure 15) since one can be obtained from

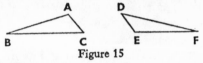

Figure 15

the other by keeping it exactly the same shape and just turning it in space (rigid motion). Euclidean geometry—the geometry that involves measurement or metric—deals with the properties of figures that remain constant, or invariant, when subjected to a particular class of transformation, viz—rigid motions.

But suppose a circle and a pair of perpendicular diameters are drawn on a sheet of rubber. If the rubber sheet is compressed, say, to half its original width, the circle will be changed to an ellipse, and the diameters of the ellipse will no longer be at right angles. But others of the original geometrical properties are not destroyed. For example, the centre of the ellipse remains at the mid-point of both diameters, as was the case in the original circle. Some spatial properties continue to exist even when the transformation is more drastic than that of rigid motion (Figure 16).

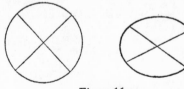

Figure 16

2. Projective transformations. A little must now be said about projective geometry. This geometry deals with the spatial properties that remain the same when subjected to

another class of transformations. These transformations are not restricted to rigid motions, but they are not so drastic as to include all possible classes of deformation. They are, in fact, intermediate in character, and are known as the class of *projective transformations*.

If an artist looks at a landscape and then records what he sees on canvas, the painting can be regarded as a projection of the landscape. On the canvas things appear as they are in vision, and not as they are in external reality. For example, no parallel lines in the landscape are recorded as parallel in the picture, since such lines appear to the eye to converge, in contrast, of course, to what we should find if we measured the distance between the landscape lines at successive points along them. In projective transformations lengths and angles are distorted to an extent depending upon the relative position of the objects drawn. Nevertheless, the geometrical structures of the original landscape can be recognized in the painting, because certain geometrical properties and relationships remain invariant under projective transformations. Projective geometry studies all such properties and relationships. A point always projects as a point, and a line in one plane always projects as a line in any other plane. Again, if four points ABCD on a straight line, project into points WXYZ on another line, the *cross ratio* $\frac{CA}{CB}\Big/\frac{DA}{DB}$ remains constant (Figure 17).

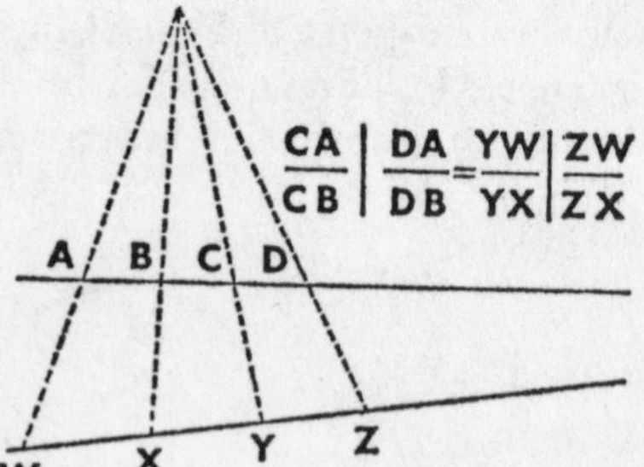

$$\frac{CA}{CB}\bigg/\frac{DA}{DB}=\frac{YW}{YX}\bigg/\frac{ZW}{ZX}$$

Figure 17

Length, angle, area, volume, etc, remain unchanged under rigid motion; point, line, cross ratio and certain other quantities remain unchanged under the wider group of projective transformations.

3 Topological transformations. Finally the group of topological transformations must be considered. These are so general and so drastic that length, angle, area, volume, point, line, cross ratio, etc are all lost; and yet certain other more fundamental geometrical properties and relationships remain invariant. The sheet of rubber on which we drew the circle can now be bent, twisted, compressed or stretched in every possible way, provided it is not torn. In fact, if figure X is subjected to a topological transformation and becomes figure X^1, two (and only two) qualifications still hold:

1 To each point p of figure X there corresponds just one point p^1 of figure X^1, and conversely.

2 If p and q are any two points on figure X, and p moves so that the distance between it and q tends to zero, the distance between the corresponding points p^1 and q^1 on figure X^1 will tend to zero, and conversely.

Topology is the study of the spatial properties and relationships that persist when figures are subjected to such drastic deformations that all metric and projective properties are lost. Some of these topological relationships will be given when we discuss Piaget's work.

It can now be seen that rigid motions and projective transformations are special cases of topological transformations.

THE VIEWS OF PIAGET AND INHELDER ON THE DEVELOPMENT OF THE CHILD'S CONCEPTION OF SPACE

Piaget and Inhelder (1956) have described the results of many interesting and ingenious experiments which led them to advance a comprehensive theory of the child's conception of space. They suggest that the child's first concepts of space are topological ones. That is, the first spatial relationships that he can represent to himself in thought, are those that deal with such characteristics of external reality as:

1 proximity or nearbyness. The young child of 4–5 years will be able to represent to himself *closeness*, before he can think of, say, *similarity*,

2 separation,

3 order, or spatial succession,

4 surrounding, or enclosure. The child, it is believed, gets the idea of insideness or outsideness before he gets any notions involving measurement of the enclosure,
5 continuity of lines and surfaces.

The above characteristics remain unchanged when the body is bent or stretched, whereas 'direction' or 'measurement' do not remain constant under these conditions.

Young children were given a number of shapes cut out of cardboard, which they had to examine by tactile exploration (feeling) without seeing them, and had then to identify each figure in turn from drawings of the shapes, the drawings being the same size as the cardboard shapes. The shapes included a circle, ellipse, square, trapezium, semi-circle with notched straight edge; also curvilinear shapes displaying various topological relationships (so the authors believe) e g having one or two holes, or open and closed rings. The authors state that the child first identifies shapes that display topological relationships. Up to 4 years of age he cannot distinguish a circle from a square, since both are closed figures, but he can identify a horseshoe shape. Later, curved edges are distinguished from straight edges, and shapes with long sides from those with short sides, so that shapes that display Euclidean relationships are progressively identified.

Children were also set to draw various Euclidean figures, e g square, ellipse, triangle, circle; also figures made up of triangles and circles, in which the components touched or intersected one another; and closed curves which had a small circle either just inside, on, or outside the curve. The authors state again that it is the closed curves with the little circle just inside or outside which are first correctly drawn, whereas the square, circle and ellipse are all drawn alike. For in their view the former display topological concepts, whereas the latter display Euclidean ones. Later, there is a progressive differentiation of Euclidean and topological relationships in the drawings.

The two investigators suggest that from about the age of 6 years topological concepts slowly give way to projective and Euclidean concepts. In projective space, it will be remembered, objects are located relative to one another although there is no measurement. Piaget and Inhelder argue that projective space begins, psychologically, when an object is no longer thought of in isolation but commences to be considered in relation to 'a point of view'. That is, the child begins to appreciate how objects appear, when viewed from different positions. In one very important experiment children were given matchsticks stuck into small amounts of plasticine and told to make a straight line. Up to 4 years of age the child, according to the authors, does not use the word 'straight', and he constructs a wavy or curved line. The sticks are placed beside one another (nearby-ness is a topological concept) but in a curved line (there is no concept of a straight line in topology). By 6 years or there-abouts the child is said to sight along the line from one end, and is said to be able, thereby, to make the line straight. Other interesting experiments involved problems of simple perspective, shadows, viewpoints, and cross sections of solids. For example, children had to select drawings that represented what a doll would see when looking at a group of three mountains from different positions. Or, they had to predict and draw the sections of certain solids, and to predict and draw their *development*, i e their unfolded surfaces. Young children could do none of these tasks. In the problem of drawing sections they understood what was required of them, but they would show a solid cut in different ways in the one drawing. When the solid was actually cut some expressed surprise, and said in effect 'You can think now because it has been cut'. But from 7 years of age onwards the sections were increasingly drawn correctly, although developments still showed errors, as when the face of a cube was drawn as a parallelogram. It is in the period 7–11 years of age that relations such as *left-right*, *before-behind* become co-ordinated so that the mental operations can be performed which enable the child to see objects from another point of view. The authors stress that the inability to make a development of a solid was due to lack of actual folding and unfolding, just as lack of experience in cutting cross sections of solids hindered them when imagining these sections. They also claim that perspective begins to appear spontaneously in drawing at about 9 years of age, while from 11–12 onwards the *development* of a cube can be drawn correctly.

We turn next to Euclidean space. In this, objects are located by means of axes of reference (length, breadth, height) and the child develops his ideas of measurement or metric, so that he can draw a Euclidean figure such as a rectangle, and can measure its sides. Once more the authors have provided some fascinating experiments. Children had among many tasks:

(i) to draw figures similar to ones given,

(ii) to forecast the 'tilt' of the water level when a glass bottle was tilted. This was said to show the development of the concept of horizontal and vertical—and hence axes of reference, through coming to use the table top as a reference,

(iii) to show their developing skill in placing objects on a model landscape where they had to bear in mind both distance and direction, thus forming a co-ordinated system.

Under 6 years of age these tasks are quite beyond children, but from 7 onwards there is increased ability to recognize similar triangles. It is not until 9 years that horizontal and vertical are well understood and cease to be found by trial and error.

There is not the slightest doubt that Piaget and Inhelder have provided many excellent experiments demonstrating that children arrive slowly at the spatial concepts which appear so matter of fact to educated adults. But it is not possible to say if their general thesis that the child's conception of space begins with topological concepts, which are transformed concurrently into concepts of projective and Euclidean space, is correct. In the first place, are the relationships which are said by the authors to be topological in fact topological using the word in the mathematical sense? Or, are young children perceiving and forming concepts involving certain relationships, which can be put into more precise terms by using topological relationships (which are essentially abstract), in Euclidean space? We can speak of two figures being topologically equivalent, but we cannot say that *a shape* displays topological relationships or properties. Secondly, Lovell (1959) has given an account of work with some 140 children between $2\frac{11}{12}$ and $5\frac{8}{12}$ years of age in which the results of a number of experiments of Piaget and Inhelder were reported. Some of his findings agree with those of Piaget and Inhelder and some do not, and he suggests that more experimental evidence is necessary before their general thesis can be accepted. For example, Lovell has shown that children can certainly identify, by feeling, shapes said by the authors to display topological relationships, more easily than they can Euclidean shapes as a whole. But Euclidean shapes with curved edges are picked out, by feeling, as easily as shapes displaying topological relationships. Indeed, there appeared little evidence suggesting that it is topological relationships as such which enable a child to identify some shapes more easily than others. It seems that it is gaps, holes, curves, points, corners, ins and outs, etc in Euclidean space that makes identification easy because the amount of 'information' conveyed is greater. Again, there was no significant difference between the drawing of topological properties and Euclidean properties if we confine our attention to Euclidean figures that have curved edges only. Further, there is evidence that children can make a straight line without 'taking aim'. Fisher and Gracey (reported by Fisher, 1965) have also called into question Piaget's views on 'topological primacy' in young children. Their work suggests that the linguistic categories available to the child affect his perceptual categories. Page (1959) also found that young children did not confuse curvilinear with rectilinear shapes as suggested by Piaget.

Other experiments carried out under the writer's direction but not published, have broadly confirmed the stages through which children pass, although there are some differences. In the matter of drawing the sections of a cylinder, prism and cone, the subjects certainly drew them with increasing accuracy from 7 years of age, but the majority of the drawings were not very good before 10 years of age. On the other hand, horizontal and vertical, and hence axes of reference, were understood by some 7-year-olds far better than one would expect from the Geneva results. Sheer experience of horizontal and vertical in everyday life, in standing up and lying down, in sailing boats on water and so forth, must have helped in developing these notions.

PIAGET'S VIEWS ON THE NATURE OF GEOMETRICAL THOUGHT

We must now discuss Piaget's views on how the child builds up his spatial concepts. Even if the theory which we have just discussed is unacceptable as it stands, his views on the importance of building spatial concepts via actions are very suggestive from the point of view of the teacher in the classroom.

For Piaget spatial relationships are not understood a *priori* by the child just because the structure of the human mind determines the thought that the mind can adopt. Neither are they due to images that have become linked according to the laws of association, nor are they passively impressed on his mind through sensation (i e incoming signals visual or tactile). Rather, the representation of space is due to the individual's activities that have taken place over many years. In the first instance, the young child acquires images through his perceptual activity, there being a close relationship between the activity he displays in perceiving spatial forms and the ability he has of evoking these figures by images. This perceptual activity consists of visual and tactile exploration. In the early stages his explorations are unorganized, as is evident in the child's poor capacity to represent to himself the figure he is looking at or feeling. In Piaget's view, because the child's early explorations are very poor, it is only topological relationships that he can represent to himself whether the exploration is done tactually or visually.

But Piaget is also of the opinion that more advanced geometrical thought is impossible for a subject who possesses only a collection of static images. The child must pass beyond the stage of imagery as a basis of representational thought, and he must be able to construct and transform spatial figures, and thus conceive a coherent system of spatial relationships. It is the actions that are actually performed on the objects or figures that bring this about. Here too, then, thought arises from the interiorization of actions performed; in short, geometrical thought is in essence a system of interiorized operations. The image, arising out of perceptual activity, certainly acquires the capacity to serve as a support to spatial reasoning; and images of spatial figures and images of the results of mental operations performed on these figures are necessary, too, for geometrical thought. But the vital element for bringing about coherent systems of geometrical thought is the operations.

For Piaget, spatial concepts result from internalized actions and not from images of things or events, or even from images of the results of actions. Arranging a series of objects in the mind is not just to imagine the objects already arranged, or to image the action of arranging them. Rather, the series must be arranged operationally—that is, by logical thought employing concepts.

As has often been said, this becomes possible because internalized actions, or thoughts, are capable of being arranged in many different ways and are reversible. So, following Piaget's views, the following kinds of activity will be of the greatest help to children:

1 actions in which objects are placed next to each other (proximity); or in series (order); actions of enclosing, tightening, loosening. These are said to tend to develop topological concepts;

2 the viewing and drawing of objects from different angles; folding and unfolding surfaces; cutting through objects to expose various sections; enlarging and reducing figures; rotating figures. These actions help to develop projective concepts;

3 drawing of similar figures; experiments involving horizontal and vertical lines and planes; measuring; the co-ordination of groups by distance and direction in model layouts. These activities help to develop Euclidean concepts.

Piaget stresses that children cannot visualize the results of the simplest actions until they have seen them performed, so that a child cannot imagine the section of a cylinder as a circle, until he has cut through, say, a cylinder of plasticine. As always for Piaget, thought can only take the place of action on the basis of the data that action itself provides. Smedslund (1963) has shown that mere observation of the horizontal level of a water surface when the container is tilted brings about no learning concerning horizontality in 5–7-year-olds who have no concept of horizontality to start with, and only limited improvements in those who have initial traces of the concept.

Further, Beilin (1966) studied the ability of 180 pupils with a mean age of 7 years 6 months to represent the water level in jars tilted at various angles. 'Unsuccessful' subjects were then trained either by showing them the water levels after their forecast (perceptual training), or by using verbal methods. While training resulted in improved performance there was no transfer to jars of a different form. While experience is of great importance in helping the child to develop his concepts of space, it must not be forgotten that genetic causes play some part. It has long been known that ability to manipulate shapes in the mind is present by 10–12 years of age, independent of measured intelligence. El Koussy pointed out in 1935 that the ability depended on the capacity of the individual to obtain, and the facility to utilize, visual spatial imagery. El Koussy's point of view has recently received a little support from the work of Stewart and Macfarlane Smith (1959) using the electroencephalograph. Piaget would certainly admit that imagery supports spatial reasoning and geometrical thought, but is not in itself sufficient.

REFERENCES

BEILIN, H, KAGAN, J and RABINOWITZ R (1966). 'Effects of Verbal and Perceptual Training on Water Level Representation'. *Child Developm*, 37, 317–329.

DODWELL, P C (1963). 'Children's Understanding of Spatial Concepts.' *Canad J Psychol*, 17, 141–161.

FISHER, G H (1965). 'Development Features of Behaviour and Perception'. *Brit J Educ Psychol*, 35, 69–78.

LOVELL, K (1959). 'A Follow-up Study of Some Aspects of the Work of Piaget and Inhelder on the Child's Conception of Space'. *Brit J Educ Psychol*, 29, 104–117.

PAGE, E I (1959). 'Haptic Perception'. *Educ Review*, 11, 115–124.

PIAGET, J and INHELDER, B (1956). *The Child's Conception of Space*. London: Routledge and Kegan Paul.

SMEDSLUND, J (1963). 'The Effect of Observation on Children's Representation of the Spatial Orientation of a Water Surface'. *J Genet Psychol*, 102, 195–201.

STEWART, C A and MACFARLANE SMITH, I (1959). 'The Alpha Rhythm, Imagery, and Spatial and Verbal Abilities'. *Durham Research Review*, 10, 272–286.

WATTS, A F (1944). *The Language and Mental Development of the Child*. London: Harrap.

CHAPTER NINE

Concepts of Length and Measurement

BEFORE children come to school they are likely to hear many expressions used by adults and older children in relation to length and measurement. For example, most children hear their mothers speak of yards of material, or—less often—their fathers speak of feet of timber, or of the distance to the station or nearby town. More frequently, however, they hear of comparisons rather than the names of actual lengths, such as 'This is longer than that', or 'That is higher than this'. These expressions are associated with many experiences ranging, maybe, from the length of nails to the height of mountains. Likewise a child hears terms like 'near' and 'far' in relation to nearby or distant towns. Again, from his play, or through watching the activities of grown-ups, he learns that a piece of string may be made shorter by cutting a piece off, or a stick made shorter by breaking it. Likewise he learns that sticks and ropes may be joined to other sticks and ropes and so made longer. Later we shall say a great deal about the view of the Geneva school regarding conceptual development in relation to length and measurement. It is sufficient to say here that it is out of these pre-school and out-of-school experiences, and out of infant school activities such as take place in the 'free choice' period, that the child comes to understand the quality of longness or length—that is, the extent from beginning to end in the spatial field. During these experiences the child moves from visual, auditory and kinaesthetic perceptions, and actions, to concepts.

In activities involving counting a child may be asked to count the number of steps he has to take to cross the classroom.

Another child will be found to take a different number of steps. Or, the lengths of short objects may be measured by the foot— the distance from heel to toe—or by the span from little finger to thumb when the hand is stretched as far as possible. From a variety of similar exercises the teacher can help her children to understand the need for a fixed unit of length for measuring purposes. Of course, mankind has had exactly this problem of establishing fixed units, and a little history of measurement is an enjoyable and stimulating piece of work for older junior pupils.

By the upper end of the infant's school the faster learners will be ready to be introduced to one of the agreed units of measurement, viz the foot. Lengths of wood or hardboard, or plain foot rulers without end pieces or sub-divisions—which can be purchased—are given to the children, and they are instructed to measure various lengths and record their answers in a notebook. In the early stages they should be set to measure the lengths of lines drawn on the blackboard or floor, or to measure the length of pieces of string, paper, etc, all of which are cut to an exact number of feet in length. Later, they can be set to measure the length of other objects in the environment to the nearest foot, so that if an object is nearly 3 feet long it is recorded as a full 3 feet. It is good, too, to let children estimate lengths before they measure, in the hope that it will lead to estimation with increased accuracy.

With experience and maturity the pupils naturally become dissatisfied with a ruler that permits measurement to a foot only, for there are so many bits and pieces left over. This is the moment to introduce the inch, and a foot stick or foot ruler with inch marks on it. At the same time have work cards available on which there are lines drawn to an exact number of inches, or lengths of string and paper similarly cut for the pupils to measure. The next step is the measurement, to the nearest inch, of objects in the environment; the children ought frequently to express their answer as, say, 1 foot 3 inches and as 15 inches, for this will help them to understand the relationship between two units used in the measurement of length. Soon they will be found to be ready for a wall scale by means of which they can measure each other's height. This is an activity that creates great interest, since personal dimensions and growth are of great consequence to most children.

Next we come to the yard and yard stick; a necessary unit when measuring longer distances. It is helpful to have some rulers divided into 3 feet with alternate sections, say, red and white, and a second set divided into 36 inches, with alternate inches of different colours. After comparing these with the whole foot, and with the 12-inch ruler previously used, the teacher should show that the yard ruler or stick is comparable with the length of her stride. By means of graded exercises similar in type to those described for feet, and feet and inches, we hope to get the child to the stage where he can measure a length as, for example 2 yards 1 foot 9 inches. The ordinary foot ruler with end pieces, and fractions of an inch up to $\frac{1}{10}$ or even $\frac{1}{16}$ inch, can be introduced when pupils are ready for it, but with the very slow learners simplified rulers may have to be used throughout the junior school.

So far, activities and experiences that presuppose that the concepts of length and measurement are possible for children have been dealt with. Have we, however, any clues as to the first beginnings of these concepts? Are there any conditions which are necessary before understanding of length can take place at all? The Geneva school led by Piaget has carried out many interesting experiments in this field to which we now turn.

THE VIEWS OF THE GENEVA SCHOOL ON THE DEVELOPMENT OF CONCEPTS RELATING TO LENGTH AND MEASUREMENT

Piaget, Inhelder, and Szeminska (1960) have outlined the views on the way in which the child comes to understand length and measurement. In one of the experiments reported early in their book they study his *spontaneous measurement*. The experimenter showed the child a tower made of twelve blocks and a little over 2 feet 6 inches high—the tower being constructed on a table. The experimenter told the child to make another tower 'the same as mine' on another table about 6 feet away, the table top being some 3 feet lower than that of the first table. There was a large screen between the model and the copy but the child was encouraged to 'go and see' the model as often as he liked. He was also given strips of paper, sticks, rulers, etc, and he was told to use them if his spontaneous efforts ceased, but he was NOT told how to use them. The following stages were observed:

(a) up to about $4\frac{1}{2}$ years of age there was visual comparison only. The child judged the second tower to be the same height as the first by stepping back and estimating height. This was done regardless of the difference in heights of the table tops;

(b) this lasted from $4\frac{1}{2}$–7 years of age roughly. At first the child might lay a long rod across the tops of the towers to make sure they were level. When he realized that the base of the towers were not at the same height, he sometimes attempted to place his tower on the same table as the model. Naturally, that was not permitted. Later, the children began to look for a measuring instrument, and some of them began using their own bodies for this purpose. For example, the span of the hands might be used, or the arms, by placing one hand on top of the model tower and the other at the base and moving over from the model to the copy, meanwhile trying to keep the hands the same distance apart. When they discovered that this procedure was unreliable, some would place, say, their shoulder against the top of the tower (a chair or stool might be used) and would mark a spot on their leg opposite the base. They would then move to the second tower to see if the heights were the same.

The authors point out that in their view this use of the body is an important step forward, for coming to regard the body as a common measure must have its origin in visual perception when the child sees the objects, and in motor acts as when he walks from the model to its copy. These perceptions and motor acts give rise to images which in turn confer a symbolic value, first on the child's own body as a measuring instrument, and later on a neutral object, e g a ruler.

(c) from 7 years of age onwards there was an increasing tendency to use some symbolic object (e g a rod) to imitate size. Very occasionally a child built a third tower by the first and carried it over to the second: this was permitted. More frequently, though, he used a rod that was exactly the *same length* as the model tower was high.

Next, the child came to use an intermediate term in an operational way (i e in the mind), this, of course, being an expression of the general logical principle that if A=B, and B=C, A=C. Children were found to take a longer rod than necessary and mark off the height of the model tower on it with a finger or by other means, so as to maintain a constant length when transposing to the copy. But, this transference is only one aspect of measurement; the other aspect which must be understood is sub-division; for only when this, too, has been grasped can a particular length of the measuring rod be given a definite value, and repeated again and again (iteration). In the final stage it was found that children could also use a rod shorter than the tower, and it was applied as often as was necessary; so that the height of the model tower was found by applying a shorter rod a number of times up the side.

For the authors, then, the concept of measurement depends upon logical thinking. The child must first grasp that the whole is composed of a number of parts added together. Second, he must understand the principles of substitution and iteration, that is the transport of the applied measure to another length, and its repeated application to this other. On this view, measurement is the synthesis of division into parts and of iteration, just as number is the synthesis of inclusion of classes and serial ordering.

In Chapter 3 of their book Piaget, Inhelder, and Szeminska studied how children judged distance and come to conserve distance. They make a clear distinction between distance and length. The latter term refers to the size of 'filled in' space, the lengths of sticks, for example. But distance refers to the linear separation of objects, or 'empty space'.

Two toy trees of the same height were placed on a table about

20 inches apart. The child was asked if they were 'near one another' or 'far apart', all direct reference to movement or length being carefully avoided. After his answer had been given, the toy trees remained in position, but a cardboard screen, higher than the trees, was placed between them. He was then asked if the trees were still 'as near' or 'as far apart' as before, and he was pressed to give reasons.

It was found that up to about 5 years of age the child considered one part of the whole distance, so that when the screen was interposed and the distance divided into two parts, the distance between the trees was said by him to be less. He could not, of course, verbalize the fact that he was confused over the whole-part relationship. But after about 5 years of age the child said that the overall distance was less, because that part of the distance between the trees taken up by the screen must be taken away since it was 'filled space'. However, after age 7 distance was conserved in spite of objects placed in between, and the notion of distance became operational.

In the next chapter of their book the authors discuss other notions about the child's understanding of distance and length. They maintain that conservation of distance, which brings about the idea of a stable and homogeneous environment, also brings the concept of the conservation of length. For, in their view, conservation of length can only be attained when the child grasps that the site occupied by an object remains the same length when the object is removed; and obversely, the size of a site that was previously empty remains precisely the same when it is filled with an object.

One prerequisite for measurement is that the child understands that an object remains the same length irrespective of change of position. To study the development of the conservation of length the children were shown two identical straight sticks, each about 2 inches long. The sticks lay parallel to one another and their ends coincided. They were asked whether the sticks were of the same length or whether one was longer than the other. All the children are stated to have judged the sticks to be of equal length. One of the sticks was then moved forward about $\frac{1}{3}$ inch and the children were questioned again about their length. Once more Piaget,

Inhelder, and Szeminska found that up to about 5 years of age children judged the stick that was ahead to be the longer, but from 7 onwards conservation was attained. However, the conservation of length when an object undergoes a mere change of position does not necessarily imply an understanding of measurement. Although a child may be able to conserve length if two sets of objects are presented in straight lines and in exact alignment, he may not conserve if one set is rearranged in a curved or zig-zag line. For example, two parallel rows of matches were presented to children, the matches being placed end to end with the rows about $\frac{1}{2}$ inch apart. The matches in one of the rows were then rearranged to form (a) a right angle, and (b) a series of zig-zags. The children were asked, for each arrangement, if the lines were of the same length; or, if an ant was walking along each line of matches, whether both ants would walk the same distance. In the younger children conservation of length was lost when the row was modified, but with increasing age conservation was present under all circumstances.

The authors next describe experiments in which a study was made of the development of a child's skill in selecting a sub-unit and in applying it. The children were asked to say which of a number of paper strips (pasted on cardboard) was the longest, which the next longest, and so forth. The strips were in the shape of right angles, acute angles, etc. After judgment had been made, they were shown a number of adjustable strips (i e the strips could be made longer or shorter) and asked in simple language to check if their estimates were correct. Later on, the children were given short strips of card of length 3 cm, 6 cm., and 9 cm (these lengths corresponded with those of the segments of the paper strips). The experimenter might then, beginning with the point of origin, apply the 3 cm card two or three times along the mounted strips, explaining that a little man was walking along the road and that these were the successive 'steps' he took as he walked. The child was asked to finish what the experimenter had begun, thus checking his estimated ranking order of length.

Once more it was found that up to about 5 years of age the idea of a unit of measurement was quite beyond the children. Somewhat later the role of a measuring unit was arrived at by trial and error, but failed to be operational. From about 7 onwards, however, measurement was understood.

Students working under the writer's direction have carried out a number of the experiments * described in the book by Piaget, Inhelder and Szeminska, with children aged 5–11 years, and a few of our findings are now given:

1 Two 'Red Indians' (A and B) were placed on a table 20 inches apart, and children were asked, individually, if the two Indians were 'near one another' or 'far apart'; all reference to movement or length being carefully avoided. Next, a piece of hardboard (with edge on table), followed by a piece of 2 inch by 1 inch baton with the 1 inch edge on table, were placed between the 'Red Indians', and each child was asked if the Indians were still 'as near' or as 'far apart' as before. The replies of the children agreed fairly well with those given by Piaget, *et al*, and the stages that they suggest seem to be there in the case of many children, although chronological age is not a very good guide to the stage of development. Thus,

   (a) J E aged 6 years could not bring the two parts together in any way when the screen or baton was interposed.

   (b) D P aged 5 years conserved the distance between the Indians when the screen was interposed, but did not maintain that the distance AB was the same as the distance BA.

   (c) W B aged 5 years could not conserve the distance, but recognized the symmetrical character of the interval AB=BA.

   (d) V P aged 8 years gave good operational explanations, such as 'The Indians are still standing in the same place', or 'They haven't moved'. On the other hand, not all the 8- and 9- year-olds were at the stage of operational thinking in this experiment.

* Thirteen experiments were each undertaken by 75 primary school children aged 5–11, and 50 E S N special school children aged 9–15. See Lovell *et al* (1962).

2 Two 10 centimetre rods, each of 1 cm by 1 cm cross section, were placed with their extremities coinciding, and most of the subjects agreed that they were the same length. The following procedures then took place, (*a*) one rod was pushed about half a centimetre ahead of the other, (*b*) the rods were placed to form a letter T, (*c*) the rods were placed at an acute angle to one another but touching, and on each occasion the children were asked if the two rods were still of the same length. Again the stages proposed by Piaget seemed to be present. At 5, 6 and 7 years of age some children failed, in all three instances, to conserve the length of the rods, others conserved in, say, (*a*) and (*b*) but not in (*c*), and so on. Some said, 'They look the same', from which it is to be supposed that they arrived at the correct answer intuitively and not operationally. On the other hand, C L aged 7 years was correct in all three instances, and said each time, 'You've only moved them'—a clear indication that his thinking was operational. But it must be borne in mind that some 10-year-olds gave incorrect answers. While the stages suggested by Piaget can be recognized in children, it must be stressed here, as elsewhere, that it is doubtful if children or adults can subject perception to logical thought until sheer experience of the situation gives some aid.

In another experiment of the Geneva school each child was required to move his bead along a wire the same distance as the experimenter moved his bead along a parallel wire. Sometimes the beads of the experimenter and child were not opposite one another to begin with, or the experimenter moved his bead from one end of his wire and the child moved his bead from the opposite end of the parallel wire. In both these instances, when the concept of measurement was not available to the child, he would, in the earliest stages, line up his bead opposite the experimenter's regardless of the distance he had moved it. Later he would move his bead by visual estimate and get the distance more or less correct. In neither instance was it possible for him to use the measuring instruments provided. At the next stage he can use his ruler to measure a length shorter than the ruler itself. But it is only in the final stage that he can measure a length longer than his ruler by the repeated application of the measuring instrument, i e by the iteration of a given fixed length. This is a very revealing experiment. Once again our experiments broadly confirm those of Piaget. The evidence from our study is clear that a child can be taught to use his ruler in rote fashion but he will not necessarily be able to decompose a length in his mind and understand what he is doing. That is not to say that we must always wait for the child to be 'ready'. When operational thinking is almost possible for the child, our teaching may well 'precipitate' the ability to measure with understanding.

REFERENCES

LOVELL, K, HEALEY, D and ROLAND, A D (1962). 'The Growth of Some Geometrical Concepts'. *Child Developm*, 33, 751–767.
PIAGET, J, INHELDER, B and SZEMINSKA, A (1960). *The Child's Conception of Geometry*. London: Routledge and Kegan Paul.
SMEDSLUND, J (1963). 'Development of Concrete Transitivity of Length in Children'. *Child Developm*, 34, 389–405.

CHAPTER TEN

Concepts Associated with Area and Volume

AREA

THE word *area* may be defined as *amount of surface*. When considering the area of a book cover or table top we have literally to spread our hands over the object and indicate the extent of its surface. The child in pre-school and early school days encounters many situations in which the amount of the surface of some body comes within his perception. He sees table tops, floors, book covers, plates, coins, the coloured tiles he makes designs with, bricks, desk tops and walls of very different surface sizes; indeed he sees a hundred and one objects that display a surface. Furthermore, he has experience of placing one thing over another and of other relevant actions. Slowly he builds up in his mind some notion of area or size of surface, although it will be a long time before he can calculate even the area of, say, a rectangle. Before the concept of area has developed to any extent the child may centre on one aspect of the surface at a time, e g its length, and say that a longer surface is also 'bigger'. Even when he has attained the concept of area he (like adults) may not verbalize his knowledge with precision; he will say, 'This table is bigger', when he means that its area is greater.

Piaget, Inhelder, and Szeminska (1960) have provided us with information that throws some light on the stages through which the child passes in developing concepts associated with area. In one experiment they attempted to find out at what age the following proposition appears axiomatic, that is, a self-evident truth: If two equal parts are taken from equal wholes, the remainders will also be equal. A child was shown two identical sheets of green cardboard approximately 12 inches by 8 inches, and told that they were meadows or fields. In the centre of each a small toy cow was placed. A small house or hut was then placed in one of the meadows and the child was asked, 'Do both cows now have the same amount of grass to eat?' It was found that every child tested said that the cow in the field without the house had more to eat. An exactly similar house was then put in the other field and most of the children agreed that there was now the same amount of grass for both cows. The authors conclude that the necessary character of the axiom seems to be present at all ages. Extra houses, all identical with one another, were then placed in each field; in the one they were placed tightly side by side as houses in a street, whereas in the other field the houses were spread about. But there was always the same number of houses in each field. As extra pairs were added the child was questioned about the amount of grass available for each cow. The experiment continued until some fifteen to twenty pairs of houses had been added. It was found that even at 5 to 6 years of age the children refused to admit that the remaining areas of the fields were equal after the first pair of houses had been put in. Between 6 and 7 years of age they would often agree that the remaining parts of the field were equal up to, perhaps, twelve pairs of houses, but with the addition of more houses the perceptual configurations became so different that the child might then deny that the remaining areas were equal. But from 7 years of age onwards the remaining amounts of grass were always recognized as equal, because the logical concept of reversibility enabled the child to carry out, in his mind, the necessary manoeuvres relative to the problem.

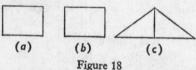

(a) (b) (c)

Figure 18

In another important experiment the child was shown two rectangles (a) and (b) which he recognized as congruent. Rectangle (a) was then cut diagonally into halves and the pieces put together to form a triangle (*c*). The question for the child was whether shapes (*b*) and (*c*) were of the same size. It was found up to 5 or 6 years of age the area was not conserved, while between 6 and 7 years of age some true judgments were made, but intuitively. From 7 years of age onwards, however, there was operational conservation of area when the shape was altered, but it was not until 8 to 9 that the child fully understood repeated application of a unit (e g a square inch) of area to calculate the amount of total surface.

Students working under the writer's direction have again repeated some of the relevant experiments described by Piaget, Inhelder and Szeminska and our findings for two of these experiments will be briefly reported.

1 The experiment involving toy cows eating 'grass' in two fields in which 'houses' were being 'built', was undertaken by a number of children between 5 and 11 years of age. All the subjects admitted that there was the same amount of grass to eat in both fields before any houses were built, and all admitted that when one house was built in each field, there remained for the cow the same amount of grass in each. But we found that most children either denied conservation of area, as soon as two houses were placed in each field (in the manner described by Piaget *et al*), or else went right through to the end admitting conservation; in our experiments up to sixteen houses in each field. Indeed, we found very few children who admitted conservation up to four, five or six pairs of houses and then denied conservation because of greater perceptual differences. Why our finding is at variance with that of Piaget *et al*, it is not possible to say. Chronological age was not a very good indicator of performance, for some children at 5 and 6 years of age conserved area right to the end of the experiment, with adequate reasons, but some older children did not. Nevertheless there was a progression in performance with age.

2 Children between 5 and 11 years of age were given (*a*) shapes made up of a number of squares and a separate unit square; (*b*) shapes made up of squares and adjoining

triangles, and both a separate unit square and a separate unit triangle such that the latter was formed from cutting a unit square along a diagonal. In both (a) and (b) the child had to find the areas of the shapes, using the units. In (b) there was also the problem of converting the number of unit squares into unit triangles, as the areas of the complete shapes could only be expressed as whole numbers in terms of the unit triangles. Our results show that there is a progression with age in the ability to iterate the unit square or unit triangle; but some 7-year-olds are better in this respect than some 9-year-olds. Generally speaking (a) was a much easier task than (b), although there were 7-year-olds who could successfully work all the exercises. It is obvious, however, that unless the child has grasped the principle of repeated application of a unit (iteration), the kind of activities that will be described in a moment will not be understood by him.

It is now possible to discuss, briefly, the kind of exercises that children can undertake in the classroom to lead them to the stage where they can calculate, with understanding, the areas of simple shapes made of squares or rectangles. One may begin by challenging our pupils (3rd or 4th year junior school) as to the best way of finding out, say, which of two sticks or classrooms is the longer. Having elicited that it is usually necessary to use a ruler, tape, etc, on which there are fixed units of length that can be iterated, the teacher may present the children with a number of pieces of paper or cardboard that are all rectangular in shape but which have slightly differing areas owing to variations in width and length. Their problem now is to find which of these has the greater amount of surface. This task necessitates the introduction of a new unit with which to cover the surface, namely the square inch. (Later one may introduce the square foot and square yard. Children should practise drawing squares of side 1 inch on scrap paper, but squares of side 1 foot or 1 yard are best drawn on the blackboard or the classroom floor or playground). A large number of squares of stiff paper or cardboard each 1 inch by 1 inch are also needed, for each pupil.

The next phase now becomes possible. Each child in a group, or whole class, is set the task of drawing a series of squares and rectangles of dimensions, say, 4 inches by 2 inches, 3 inches by 3 inches, etc; and of dividing these shapes into 1 inch squares. Each shape, in turn, is then covered with the 1 square inch squares of paper or cardboard, and each child has to discover the number of squares required to cover each of his shapes. Instead of the pupils having to count up the number of squares each time to find the total area, they must be helped to discover the generalization that the area of a square or rectangle is equal to the number of units of length multiplied by the number of the same units of breadth. Similar exercises can be carried out on the floor using squares of hardboard of side 1 foot to cover a rectangle or square chalked on the floor and divided into square feet. Thus, one may now define the area of a surface as its measurement in square units. Note carefully that the definition does not specify whether the units are square inches, square feet, or square yards.

It is helpful at this stage, too, to get each child to make a square out of, say, 36 squares each 1 inch by 1 inch, and to rearrange the square into rectangles of varying dimensions, e g 12 inches × 3 inches, 9 inches × 4 inches. This exercise helps the child to develop operationally the conservation of area, and it provides another opportunity to point out that the area of a shape which is not a square is nevertheless composed of a number of square units. Furthermore, one can indicate that the square is a special case of the larger family of rectilinear figures.

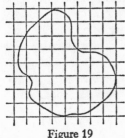

Figure 19

The possibility of the area of an irregular figure being composed of a number of square units, e g square inches, is a difficulty that faces many secondary school pupils, and one they find hard to understand. Some pupils can be helped by drawing an irregular shape on square paper and getting them to shade in, and count, the total number of whole squares enclosed. It can then be explained that the *parts* of squares that are enclosed could be rearranged into a rectangle or square and this in turn would have an area expressed in square units. If the pupil has not sufficient mental manoeuvrability to be able to deal with the 'bits and pieces of squares' in his mind in the fashion indicated, there is little one can do to get him to understand. The problem of the area of an irregular-shaped figure must be left until later if he is to comprehend what he is about.

VOLUME

The word *volume* may be defined as the *amount of space*. In this instance we have:

1 to move our hands about inside, say, a box or cupboard, in order to indicate the amount of space within;
2 to pass our hands around a box, block, or ball, to indicate the amount of space taken up by the object.

In both cases, and in all examples illustrating these cases, one must make clear that three dimensions are involved. Later, we shall consider some practical teaching points relevant to the calculation of the volume of cubes and cuboids, but first it is necessary to discuss the views of the Geneva school on the development of concepts associated with volume.

Piaget, Inhelder, and Szeminska (1960) have performed experiments which throw light on the development of the concept of volume. Two of their experiments will be described a little later, but we must first revert for a moment to the development of the concept of conservation of substance. It has already been shown that the junior school child is increasingly able to conserve 'amount of substance' with age and experience. But, according to the authors, the child of, say, 9 years of age who is able to conserve the quantity of plasticine when its shape is changed, may still think that the substance might have expanded or contracted. The child will not use these exact words, but if a ball of plasticine is immersed in water after having been elongated into a 'sausage', he will likely say that the 'sausage' will 'take up more space' or 'less space' than before. It is not until twelve that, in their view, the volume of the plasticine is conserved.

In one experiment children were shown a solid block, representing a house, 4 cm high and on a base measuring 3 cm by

3 cm. Other cardboard bases were also provided, measuring, say, 2 cm by 2 cm, 3 cm by 4 cm, 1 cm by 3 cm, etc, and the children were told that they had to build another 'house' with 'as much room' on one or more of these latter pieces of cardboard. They had, of course, to keep the volume constant while altering the area of the base on which the house was built, and instructions had often to be given to the effect that the house must be kept on the cardboard base and must not extend beyond it. Moreover, each new 'house' had to be built out of cubes 1 cm by 1 cm by 1 cm.

In a variation of this experiment the child was asked to make his new 'house' alongside the model. The authors claim that in this instance they frequently found that he built walls all round the original model on five or even six sides, so that he tried to reproduce, not the solid object, but the volume of that object in so far as it consisted of so much empty space bounded by the surfaces. This behaviour, in the authors' view, is most significant; for they believe that it demonstrates that the child's first notions of volume are topological in character, that is, the child first thinks of volume in terms of boundary surfaces rather than of space occupied. In their view, Euclidean characteristics of volume will not be understood until much later; this, likewise, explains why calculations involving volume are not understood until about 12 years of age.

A second experiment consisted in showing the child a set of metal cubes of side 1 cm, which were then placed at the bottom of a vessel of water. The experimenter built the thirty-six unit cubes into a block measuring 3 cm by 3 cm by 4 cm and the child noted the extent to which the water rose in the vessel. The child was then asked if the water level would change if the unit cubes were rearranged to make a block, say, 2 cm by 2 cm by 9 cm. This experiment enabled the child to be questioned about three different meanings that can be attached to the word volume, viz:

1 internal volume, or conservation of the unit cubes;
2 volume as 'occupied' space, or the amount of 'room' that the 36 cubes took up in the water;
3 complementary volume or the amount of water displaced.

The authors maintain that if the child understands the second and third meanings, he will anticipate that the level of the water remains the same.

The findings of Piaget, Inhelder, and Szeminska are summarized as follows: at about 5–6 years of age the child made his 'house' by paying attention to only one dimension of the model, usually its longest dimension. The 'house' was usually made the same height as the model, regardless of the fact that the base of the 'house' was smaller than that of the model. Further, when reproducing a given volume he tended to follow bounding surfaces. Between 6 and 7, however, there was an attempt to make an equivalent volume on a different size base, and the 'house' was often made taller than the model.

Between 7 and 8 or 9 new developments were found. The child still did not measure and make exact correspondences in dimensions by using unit cubes as units, and by finding out how many cubes were necessary to equal a side of the model in length. Nevertheless, he was often successful in copying, exactly, two of the three dimensions of the model. As far as conservation of volume was concerned he could only understand interior volume, or the invariance of the amount of matter contained within the surfaces (viz, the number of unit cubes at the bottom of the glass vessel). There was still no grasp of 'occupied' volume, so that when the cubes were rearranged in the vessel they did not, in the child's view, necessarily take up the same amount of room as before. Then, starting about 8 to 9 years of age, the children began to use the unit cubes with which to measure the dimensions of the model, but they still could not establish any numerical relationships between lengths, areas, and volumes.

Finally, beginning at 11 to 12 years of age, the child came to understand that the volume of the block and the 'house' he built were equal if their respective dimensions (i e their lengths, breadths, and heights) were the same. Furthermore, he acquired the notion of 'volume occupied' by an object compared with the 'interior volume' of the object defined by its bounding surface.

Lunzer (1960) has made a follow-up study of the development

of the concept of volume as suggested by Piaget, Inhelder, and Szeminska. First, he summarized as follows, the stages proposed by the authors:

1   Below an age of 6–6½ children have no understanding at all of conservation of volume.
2   During the junior school stage there is increasing understanding that a solid composed of a number of units can be altered in shape without any change in the amount of 'room' in the 'house'.
3   For a period some young children think of volume in terms of what is surrounded by boundary surfaces.
4   At or about eleven to twelve years of age children discover:

   (a) the method of calculating volume as a function of length, breadth, and height;
   (b) the notions of infinity and continuity;
   (c) the conservation of displacement volume; that is, the amount of air or water displaced by an object.

5   The three developments noted under the last heading are all interdependent in so far as 4(a) and 4(c) depend upon (4b).

Using groups of children aged from 6 to 14 years he reported that 1, 2, 4(a) and 4(c) were completely borne out, but there was no evidence for 3 and 5. Nor was there any indication that 4(b) was of any real significance to the subjects. Lunzer points out that 1 and 2 represent just another instance of the growth of conservation that takes place between 6 and 8 years of age, and he suggests that conservation of displacement volume is something that can be spontaneously recognized at a certain stage of development. But, because comparison of displacements (e g watching the heights of the water in a glass vessel) so rarely occurs in the everyday experience of the child, and because conservation of displacement volume must depend on the prior recognition of the conservation of the volume of water that is displaced, this type of conservation inevitably comes later. Further, he points out that it is doubtful if the operation of multiplying three linear measurements together to calculate the volume of a cuboid would appear spontaneously to the child, and thinks its appearance is due to its being taught in school. Lunzer thinks that 3 and 5 are necessary for Piaget's theory since it demands the priority of topological relationships over Euclidean. Lunzer thus casts doubt on a theory that has already been called in question by Lovell (1959).

A study of the development of the concept of *physical* volume in junior school children has also been carried out under the writer's direction (Lovell and Ogilvie 1961). Notions of internal volume, volume as 'occupied' space, and complementary or displacement volume were studied in a group of 51 first year, 40 second year, 45 third year, and 55 fourth year children, making 191 children in all. The apparatus consisted of a number of plastic cubes of side $\frac{7}{8}$ inch, a gallon can and a pint can. All children were suitably and individually questioned, and the major steps in the testing programme are given below:

1 two blocks were presented. In one the cubes were arranged as $2 \times 2 \times 3$ and the other as $2 \times 3 \times 2$. One block was then changed to the arrangement $1 \times 2 \times 6$. Pupils were questioned about the conservation of volume.

2 Each pupil was shown the empty gallon can. A block, $2 \times 3 \times 2$, was placed in the gallon can, and the child asked if the same amount of water could still be poured into it. The cubes were then rearranged as $1 \times 2 \times 6$ and the question repeated.

3 The gallon can was presented empty, and the pint can presented full of water. Each child was asked what would happen to the water in the pint can if (a) a block $2 \times 3 \times 2$, and (b) a block $1 \times 2 \times 6$ were lowered into the pint can very carefully so that there was no splashing of water.

4 Each subject was asked what would happen to the water if just one cube was lowered very carefully into (a) the full pint can, and (b) the full gallon can.

5 Children were asked to compare the amount of water spilled if (a) a cube of wood, and (b) a cube of exactly the same size and shape but made of lead, and therefore much heavier, were lowered very carefully into the full pint can.

6 The subjects had to compare the amounts of water spilled by a single cube when placed at the bottom of a full can,

with the amount spilled when suspended by a fine thread half-way down in the can.

The results may be briefly summarized as follows:

(a) some two thirds of the first and second year pupils, and over 90 per cent of the third and fourth year pupils, conserved the internal volume of a block made up of twelve cubes. It is possible that the figures would not have been as high if more cubes had been used.

(b) in 2 above, 40 per cent of the first year pupils, and over 80 per cent of the fourth year pupils conserved 'occupied' volume.

(c) over 90 per cent of children in each age group maintained that water would be displaced if a block was lowered into the pint jug when full of water. But only some 55 per cent of the first year pupils, and 75 per cent of the fourth year pupils maintained that the amounts of water displaced were the same when the $2 \times 3 \times 2$ block was replaced by the $1 \times 2 \times 6$ block.

(d) a number of the younger children who maintained that water would be displaced if a block of twelve cubes was lowered into full pint can, denied that water would spill over if a single cube was lowered into a full gallon can. The notion of displacement volume becomes more generalized with age so that the amount of water spilled, for a given volume of cubes, is independent of the size of the container.

(e) many children think that a heavy cube will displace more water than a lighter cube of the same size and shape.

(f) in a number of children the amount of water displaced from a full can is thought to change according to whether the cube lies at the bottom of the can, or is suspended in the water and completely immersed.

(g) for junior school children, displacement volume often seems to be dependent upon the weight of the object immersed, depth to which it is immersed, size of container and other factors. The irrelevant influences are eliminated only slowly with age.

(*h*) if we neglect the answers to the question about the amount of water displaced in relation to the depth of the the cube in the water, No 6 above, it was found that only 3 pupils in the first year, 5 in the second, 5 in the third, and 21 in the fourth years answered all the other questions correctly.

It is clear that an understanding of physical volume in any generalized sense does not develop until late in the life of the junior school child, and even then there are many gaps in his knowledge. Children have to learn to eliminate the irrelevant factors. This is a slow business. The junior school child finds it very difficult, if not impossible, to consider the effect of one variable, holding the other variables constant.

It is possible, but not certain, that children could learn more quickly about physical volume by being exposed, in school, to learning situations where the effectiveness of the relevant and non-effectiveness of the irrelevant variables could be made evident in the same experiment. It seems that Piaget is optimistic if he thinks that the *single* test in which he employed thirty-six unit cubes in a bowl, will enable him to distinguish between those who have developed a complete concept of physical volume from those who have not. Again, our evidence does not deal with the problem of volume from the mathematical point of view. If the child's ability to calculate volume parallels his ability to understand physical volume it may be due only to general intellectual growth. It may be possible to measure the space taken up by a cube or cuboid before the concept of physical volume has become well developed.

Goodnow (1962) has reported an interesting cross cultural study in which she investigated the growth of the conservation of volume, the conservation of area (the experiment involving the toy cow in a field), and the conservation of weight. The children involved were 10, 11, 12 and 13 years of age in Hong Kong and were drawn from very different socio-economic backgrounds. She studied Europeans, Chinese from high-ranking Anglo-Chinese schools, Chinese schoolboys of low socio-economic status, and Chinese boys with low socio-economic status and 'semi-schooling'. On the conservation tasks, milieu, schooling and socio-economic status had far less effect than had been anticipated although the various groups were far apart on the Raven's Progressive Matrices test. At the same time, the tests of conservation of area and volume were harder for all the Hong Kong groups than had been expected. Overall, however, the replication of the Geneva results was fairly good.

In a study of American children between 6 and 12 years of age, Uzgiris (1964) showed that conservation of substance, weight and volume developed in that order using four very different kinds of materials. Inversion of the conservation sequence was rare. But the materials used did affect the age at which conservation appeared, and it seems that the relationship between materials and conservation may be most in evidence during the formation of conservation schemas.

We can now turn to discuss more specifically an approach to the calculation of the volume of cubes and cuboids. This can be undertaken with pupils in the top class of the primary school or in the early forms of the secondary school, depending on their ability. A useful approach is to present pupils with two or more boxes, rectangular in shape, but with slightly different volumes. The problem is to find the box which has the greater amount of space inside it, i e which occupies more space. Reference may be made to the fact that a new unit was needed when area was first calculated, and this may be used as a lead up to another new unit now required, namely, the cubic inch.

We must also introduce the cube of side 1 foot and the cube of side 1 yard. The former is represented approximately by a biscuit tin, or it may be made exactly from hardboard. The cube of side 1 yard may also be made by covering a wooden framework with hardboard. As for cubes of side 1 inch, it is best to buy these made of solid wood. At least sixty-four are required, so that a number of cubes and cuboids can be built up in turn from them. If it is intended that the whole class be set to work on the problem at the same time rather than a small group, several sets of sixty-four cubes will be required.

Pupils build up the inch cubes into larger cubes and cuboids with dimensions, say, 12 inches × 3 inches × 1 inch, 4 inches × 4 inches × 4 inches, 5 inches × 5 inches × 3 inches, etc. After

I

having counted the number of inch cubes that went into the building of the solids, and having measured their dimensions, the pupils must be helped by appropriate discussion, to make the following generalizations:

1 the volume of a cube or cuboid is equal to the number of units of its length multiplied by the number of units of breadth, multiplied by the number of units of height; all units being of the same kind;

2 the volume of a cube or cuboid is equal to the area of cross section multiplied by the third dimension, with appropriate units.

A given number of cubes that have made up a solid cube or cuboid can be rearranged in various ways to emphasize:

(a) conservation of volume;

(b) a shape which is neither a cube nor cuboid still has a volume which is expressible in terms of cubes.

Hence we may say that the volume of a body is its measurement in cubes.

REFERENCES

BEILIN, H and FRANKLIN, I C (1962). 'Logical Operations in Area and Length Measurement: Age and Training Effects'. *Child Developm*, 33, 607–618.

GOODNOW, J J (1962). 'A Test of Milieu Effects with some of Piaget's Tasks'. *Psychol Monogr*, Vol 76, No 36.

LOVELL, K and OGILVIE, E (1961). 'The Growth of the Concept of Volume in Junior School Children'. *Journal of Child Psychology and Psychiatry*, 2, 118–126.

LOVELL, K, HEALEY, D and ROLAND, A D (1962). 'The Growth of Some Geometrical Concepts'. *Child Developm*, 33, 751–767.

LUNZER, E R (1960). 'Some Points of Piagetian Theory in the Light of Experimental Criticism'. *Journal of Child Psychology and Psychiatry*, 1, 191–202.

PIAGET, J, INHELDER, B and SZEMINSKA, A (1960). *The Child's Conception of Geometry*. London: Routledge and Kegan Paul.

UZGIRIS, I C (1964). 'Situational Generality of Conservation'. *Child Developm*, 35, 831–841.

CHAPTER ELEVEN

The Number System*

Two good reasons have already been given for including this chapter in the book. A third reason is linked with the second of those given in the preface. It seems likely that for the more able children (certainly for *all* teachers), somewhat greater stress will be laid in the future on the abstract point of view in mathematics with some explicit attention to axiomatics and mathematical systems. *The Report of the Commission on Mathematics* published by the College Entrance Examination Board in the U.S.A. in 1959 recommends that a study of the number system (and many other topics) should begin in Grade 9 (15 years of age) for all potential 'college educable' pupils, now some 40 per cent of the 18-year-old age group in the U.S.A. Remember we shall deal only with what are constructions of the mind.

THE PROBLEM OF THE BASE

EARLIER it was stated that the Hindu-Arabic system of representing the natural numbers is now nearly always used. It employs ten symbols, namely 1, 2, 3, 4, 5, 6, 7, 8, 9, 0; and the fact that ten are used probably derives from man's early use of his fingers in counting. This decimal system is said to have *base* 10. But it is not necessary to have 10 symbols or base 10. We could use, say, 8 as a base and then we should employ only the symbols 1, 2, 3, 4, 5, 6, 7, 0. The sequence of natural numbers would then be written as:

1, 2, 3, 4, 5, 6, 7, 10, 11, 12, 13, 14, 15, 16, 17, 20, 21, etc.

---

* Readers are reminded that reasons for including this chapter have been given in the Preface. If these reasons are ignored, the chapter may seem out of place in the book.

The notation 635 in this system indicates 5 units plus 3 eights, plus 6 of the squares of eight. Thus 635, using base 8, $=6 \times 8^2 + 3 \times 8 + 5 = 415$ in base 10. The apparatus devised by Dienes gives first year junior school children experience of working addition and subtraction of number to various bases, as has been seen.

It might be thought that the discussion of number bases is a matter of academic interest only. That is not so. In electronic digital computors much use is made of base 2, and it has been argued that base 12 is a more practical base than 10. But those are not the reasons for raising the problem here. Our discussion is aimed at making the point that there is a distinction between the numeral or symbol or figure—whatever it is called—and the number it signifies. The same numeral, symbol, or figure indicates different number-values according to its position, and according to the base in use. For example, the symbol 3 in 368 has a value other than the symbol 3 has in 263; and the symbol 13 represents the value 13 using base ten, and the value 11 using base 8. Note carefully, too, that 0 is referred to as a symbol and not as a natural number.

RULES OF ARITHMETIC FOR THE NATURAL NUMBERS (POSITIVE INTEGERS)

It is now proposed to discuss more formally the rules of arithmetic for the natural numbers, or, as they are called, the positive integers. We could also say that we are going to choose the *assumptions* which we wish the natural numbers to obey. These rules will be recognized as true for positive whole numbers but they will be formalized and put into a logical context. In the following rules a and b are any two natural numbers.

1   Closure Law for addition. The sum of every pair of positive integers is *defined* as a unique positive integer. $a+b$ is a unique positive integer; e g $2+3=5$.

2   Closure Law for multiplication. The product of every pair of positive integers is *defined* as a unique positive integer. $a \times b$ is a unique positive integer; e g $3 \times 4 = 12$.

3   Commutative Law for addition. This states that $a+b$

$=b+a$. Everyone is familiar with the fact that $3+6$ $=6+3$. In other words, order in addition is unimportant, and addition is commutative.

4 Associative Law for addition. This states that $a+(b+c)$ $=(a+b)+c$. Now addition was *defined* in the Closure Law for addition for *pairs* of natural numbers, not for *triples*. In adding $(2+4)+3$, one can, by the Closure Law for addition, add $2+4$ to give 6 and then add 6 to 3. But one could equally well have added $4+3=7$ and then added under the Law $7+2$. Addition is associative.

5 Commutative Law for Multiplication. This states that $a \times b = b \times a$. All educated people know that $3 \times 5 = 5 \times 3$, so the order in multiplication is of no consequence. Multiplication, like addition, is commutative.

6 Associative Law for Multiplication. This gives $(a \times b)$ $\times c = a \times (b \times c)$. Under the Closure Law for multiplication we know that $3 \times 4 = 12$. We can then calculate 12×5 and obtain 60. Equally well we could calculate 4×5 and get 20, and multiply 20 and 3 together and get 60. So $(3 \times 4) \times 5 = 3 \times (4 \times 5) = 60$. Multiplication, like addition, is also associative.

7 Distributive Law. This states that $a \times (b+c) = (a \times b)$ $+(a \times c)$. This law is well recognized; for example $3 \times (2+8) = (3 \times 2) + (3 \times 8) = 30$. In everyday arithmetic the law is frequently involved, although often unrecognized. For example:

$$17 \times 39 = 17(30+9) = (17 \times 30) + (17 \times 9) = 510 + 153 = 663$$

The above expressions show precisely what we are doing when we work the example as

$$
\begin{array}{r}
17 \\
\times 39 \\
\hline
510 \\
153 \\
\hline
663
\end{array}
$$

8 Identity Law for Multiplication. The number 1 is called the identity element in the multiplication of natural

numbers, because for any number a, $a \times 1 = 1 \times a = a$. This is called the Identity Law for multiplication, since multiplication by 1 leaves a identically the same as it was before.

The above eight laws are the ordinary laws of arithmetic, which most people recognize they have been using in everyday calculations, when their attention is drawn to them. It will now be shown that there are other classes of numbers (viz, numbers other than the natural numbers) in the overall number system, and each time a new class of number is introduced the laws must be shown to hold all over again. Moreover, we shall classify numbers in two separate ways; that is, we shall use two different conceptual frameworks inside which we can discuss numbers.

CLASSIFICATION OF NUMBERS—A

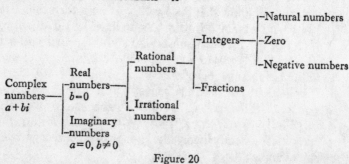

Figure 20

ZERO AND NEGATIVE INTEGERS

When using natural numbers one can always perform addition and multiplication. But we can only *meaningfully* perform, with natural numbers, the operation $a - b$ or subtraction, when $a > b$. If $a < b$, then $a - b$ has no meaning if we are confined to positive integers. A step on the way to removing this restriction was taken by introducing the symbol 0 so that $a - a = 0$. The symbol 0 stands for a new number, zero,* which is an integer, but not a positive integer or natural number (Figure 20).

* The symbol 0 also indicates place value.

It was an even greater step forward in removing the restriction when the symbols -1, -2, -3 etc were introduced, together with the definition $a-b=-(b-a)$ for the case when $a<b$. A new type of number was then brought into being which allowed subtraction without any restriction whatever. These new numbers are called negative integers and, as can be seen from the Figure 20, natural numbers, zero and negative integers are sub-classes of the class of integers. These negative integers do not correspond to anything in reality as positive integers do, but they often play a vital part in the solution of real life problems if mathematics is involved at all.

Now, having introduced negative integers into the number system, one must *define* operations with them in such a way that the original laws of arithmetic hold true. For example, to maintain the truth of the distributive law one must have $(-1)(-1)=1$. For, if $(-1)(-1)=(-1)$ then on letting $a=-1$, $b=1$, $c=-1$ we have:

$$a(b+c)=ab+ac$$

or

$$(-1)(1-1)=-1-1$$
$$=-2$$

Whereas in fact we have zero. This rule, and other 'rules of signs' that are often so troublesome for school-children, together with certain definitions that govern the operations of negative integers and fractions, cannot be proved. They are *created* by man to allow freedom of operations while preserving the fundamental laws of arithmetic.

FRACTIONS

One can look upon the symbol $\frac{3}{4}$ in a number of ways. First, it can be regarded as indicating a whole-part relationship. It is in this way that we always introduce the idea of a fraction to children, as, for example, when one cuts a piece of string into halves or a sheet of paper into quarters. Second, one can think of it as signifying a sub-unit, as, for example, when we measure in, say, half inches. Third, and most useful, we can consider a fraction as an *ordered* pair of integers. The word *ordered* is used because order is important.

As man used certain measures (e g yard, pound, etc and the corresponding units used in earlier centuries and in other countries) in everyday life he often found that the quantity measured contained a certain whole number of units and something left over. This necessitated the introduction of sub-units. For example, if the original unit of 1 is divided into k sub-units, and a given quantity is found to contain l sub-units, the measurement of the quantity is given by $\frac{l}{k}$. That is what we know as a fraction, whether a proper or improper one. It took man a long time to divorce the symbol $\frac{l}{k}$ from concrete situations involving measurement, and to look upon these fractions or ratios as pure numbers. If l and k are natural numbers the symbol $\frac{l}{k}$ stands for what is known as a rational number, and as can be seen from the Figure 20, integers and fractions make up the total class of rational numbers.

Once fractions are acknowledged as numbers there are no longer any obstacles to division. The quotient $x = \frac{l}{k}$ of two integers l and k *defined* by the equation $xk = l$, is itself an integer *only* if k is a factor of l. But if it is not we need no longer be concerned, for the symbol $\frac{l}{k}$ now represents a number just as it would have done had k been a factor of l. In one instance $\frac{l}{k}$ is a fractional number, in the other instance, an integer. In both instances $\frac{l}{k}$ is a rational number. Fractional numbers make all division possible *except* division by zero. This point will be discussed again later.

RATIONAL NUMBERS

Rational numbers consist of all the integers plus all fractions of the form $\frac{X}{Y}$ where X and Y are integers and Y is not equal to zero. But there is a snag here, since a fractional number has many symbolic representatives. For example,

$$\frac{1}{2} = \frac{2}{4} = \frac{4}{8} = \frac{6}{12} \text{ etc.}$$

Therefore a further definition is necessary, namely, that the natural number $\frac{X}{Y}$ is identical with the rational number $\frac{XZ}{YZ}$ provided $Z \neq 0$. It is now seen that $\frac{1}{2}$, $\frac{2}{4}$, $\frac{4}{8}$ etc are different names for the same rational number. Very often one makes the rational number as simple as possible, as when one 'reduces to its lowest terms'.

REAL NUMBERS

Reference to the Figure 20 shows that *rational* and *irrational* numbers together make up the class of *real* numbers. The former have already been discussed to some extent, but in order to distinguish them from the latter, rational numbers must be looked at again from another point of view. Any rational number can be stated exactly as a decimal. Thus:

$\frac{1}{2} = 0.5000\ldots\ldots$ $\frac{3}{4} = 0.7500\ldots\ldots$ $1\frac{1}{4} = 1.2500\ldots\ldots$

$\frac{1}{3} = 0.333\ldots\ldots$ $2\frac{1}{7} = 2.142857142857\ldots\ldots$

In the case of rational numbers the digits after a certain point repeat themselves, either singly as in the case of 0 and 3 in the first four examples, or in groups as in the case of $2\frac{1}{7}$. A horizontal line called a bar is usually placed over the digit or digits that repeat, hence $0.5\overline{0}$, or $2.\overline{142857}$. Thus a rational number is identical with a certain repeating decimal.

But there are other numbers that can only be represented by a non-repeating decimal. However many places of decimals one works out there is no repetition of any digit or digits. Such a number is said to be an irrational. Two very well-known examples are π and $\sqrt{2}$.

RULES OF ARITHMETIC FOR REAL NUMBERS

Earlier there was a discussion of the eight rules of arithmetic which held true for rational numbers. It can be shown that all these rules hold for all real numbers. Thus the last rule, No 8, now formally reads: The real number 1 is called the identity element in the multiplication of real numbers since any real number multiplied by 1 leaves it identically as it was before.

One can now enumerate three further laws:

9 Identity Law for addition. The real number 0 is *defined* as the identity element in the addition of real numbers, i e $a+0=0+a=a$. The addition of the real number 0 leaves a identically as it was before.

10 Inverse Law for addition. The real number $(-a)$ is *defined* as the additive inverse of the real number a. That is $a+(-a)=(-a)+a=0$. The addition of $(-a)$ undoes the operation of adding a and is said to be the inverse operation.

11 Inverse Law for multiplication. The real number $\frac{1}{a}$ (providing a is not zero) is called the multiplicative inverse of the real number a. That is $a \times \left(\frac{1}{a}\right) = \left(\frac{1}{a}\right) \times a = 1$ providing $a \neq 0$. The operation of multiplying by $\left(\frac{1}{a}\right)$ undoes the operation of multiplying by a and is said to be the inverse operation.

It can now be seen that the natural numbers are special instances of real numbers. As special instances they do not satisfy all the eleven rules that have been listed. Indeed, they cannot satisfy rules 9, 10 and 11, for 9 involves zero, 10 involves negative integers, and 11 involves fractions.

On the other hand, the natural numbers have some properties not shared by all the real numbers. For example, natural numbers can be expressed as a product of prime factors in a way that is unique except for the order of the factors. With natural numbers, too, one can use a process known as 'Mathematical Induction' to prove important theorems.

The laws that were originally true for natural numbers have

now been extended to cover the whole domain of real numbers. *This extension of the rules from one domain to a much larger domain is one aspect of the characteristic mathematical process of generalization.* The rules of addition, subtraction, multiplication, division can now be performed without restriction in the domain of real numbers. Mathematically speaking, the real numbers form a *field*.

SOME DIFFICULTIES ASSOCIATED WITH ZERO

Children frequently make errors in operations that involve zero. Here it is only possible to mention the problem of fractional numbers that have zero either in the numerator or denominator. If x and y are real numbers other than zero it might be thought that we could have $\frac{0}{x}, \frac{y}{0}, \frac{0}{0}$.

Suppose now that $\frac{x}{y} = z$, then $x = yz$. Thus by analogy if $\frac{0}{x} = w$, then $0 = wx$. But $x \neq 0$, therefore $w = 0$. Thus $\frac{0}{x} = 0$.

If $\frac{y}{0} = w$, then $y = w \times 0$ which is 0 for all values of w, and hence cannot equal y since $y \neq 0$. Hence $\frac{y}{0}$ is meaningless.

Finally, if $\frac{0}{0} = w$, then $0 = 0 \times w$. But this is true of whatever the value of w may be. Hence $\frac{0}{0}$ is indeterminate. Zero may never appear in the denominator of a fraction, but $\frac{0}{x} = 0$ for $x \neq 0$. Hence we arrive at the common saying 'Never divide by zero'.

COMPLEX NUMBERS

It will be seen from Figure 20 that to the extreme left are *complex numbers*, of which a very brief explanation will now be given. It was found as mathematics developed that there were many problems that could not be solved by real numbers alone. Earlier it was seen that to obtain greater freedom in calculation, negative and rational numbers had to be introduced. So, when it was found, for example, that a simple equation like $x^2 = -1$ had no real solution (since the square of any real number is always positive) a new class of numbers had to be brought into being, otherwise it would have been necessary to admit that such an equation was insoluble. Hence one finds the symbol i being introduced defining $i^2 = -1$. This symbol i has nothing whatever to do with counting, and it seemed so far removed from reality to mathematicians of the last century that it was said to signify an *imaginary unit*. Numbers represented by the symbols $x + yi$, where x and y are real, are called *complex numbers*.

It must be stressed that this extension to the number system must not disturb the validity of the rules and properties of other classes of numbers. Definitions associated with these extensions had to be made in such a way that the earlier laws still held. These extensions were of the greatest importance in dealing with real life problems, and their practical values provided the motivation which lead men on to extend the number domain in this way, although these values in themselves did not provide logical proof of the validity of the extension.

With this introduction it is possible to proceed to list certain definitions for addition and multiplication.

1 If x and y are real numbers, the symbol $x + yi$ signifies a complex number, which has a real part x and an imaginary part yi.

2 In addition of complex numbers:

$$(u + vi) + (x + yi) = (u + x) + (v + y)i$$

3 In multiplication of complex numbers:

$$(u + vi)(x + yi) = (ux - vy) + (uy + vx)i$$

As a particular instance of definition 3 we have

$$(x + yi)(x - yi) = x^2 - xyi + xyi - y^2 i^2 = x^2 + y^2$$

On the basis of these definitions it can be shown that the commutative, associative and distributive laws hold for complex numbers. Furthermore, subtraction and division of complex numbers lead into numbers of the form $x+yi$ so that complex numbers, too, form a field. Again, division by $0+0i$ is excluded.

CONCLUSION

In the case of an equation of *any* degree n with real or complex coefficients, as for example.

$$a_n x^n + a_{n-1} x^{n-1} + a_{n-2} x^{n-2} + \ldots a_1 x + a_0 = 0$$

then its solutions lie in the field of complex numbers, and no new classes of numbers are needed.

It is not our intention to develop the number system any further. It might be noted, however, that for certain purposes, mathematicians have developed other number systems.

CLASSIFICATION OF NUMBERS—B

Having looked at the number system in one way, we are now going to use another conceptual framework inside which we can discuss numbers. This latter framework would be regarded by many mathematicians as the best way of handling the number system.

First, we shall represent numbers by length on a straight line. Take points 0 and 1 so that 0 is to the left of 1; then the segment $(0, 1)$ is said to have length 1 by definition of length.

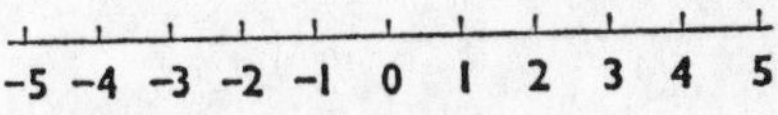

Figure 21

By sliding this segment along the line we can represent any natural number by lengths terminating at points 2, 3, 4, etc. Moreover, any number such as $\frac{1}{x}$ (where x is a natural number) can be found by dividing the segment $(0, 1)$ into x equal parts (Figure 21). Further, by stepping along, approximately, the segment $\left(0, \frac{1}{x}\right)$ we can represent, say, the number $\frac{W}{x}$ where W is also any natural number. An irrational number may be

141

placed along the line by working out the number to one or two places of decimals. Likewise by moving to the left of *0* we can represent negative integers, etc. We have now reached the stage where any real number can be represented by lengths on a line.

At this point we shall abandon the conceptual framework whereby we regarded numbers as natural, negative, real, irrational, etc and think of *directed* numbers for a while.

Take O as a fixed origin and let OX be a fixed line through O such that lengths in the direction $\overrightarrow{OX}$ are said to be positive and lengths in the direction $\overrightarrow{XO}$ negative. Now suppose we have a line OA, of *positive length*, *r*, lying along OX. Next, let the line OA be rotated through an angle θ, so that θ is taken as positive when OA rotates from OX in an anticlockwise direction and negative when it is rotated in the opposite direction (Figure 22).

The positive length *r* in the direction OA is defined as the number *r* in the direction θ, and all such numbers are classified as directed numbers.

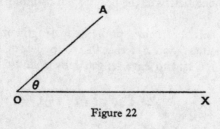

Figure 22

The directed number *(3, 0)* is, of course, simply denoted by *3*, and the directed number *(2, π)* by *−2* Further, the directed number $\left(1, \dfrac{\pi}{2}\right)$ is denoted by the symbol *i*. Strictly speaking it is the vector * joining the origin *0* and the point *i* that represents the number $\left(1, \dfrac{\pi}{2}\right)$, but a point is

---

* There are vector quantities of different types, physical and geometrical. But they all have the characteristics of magnitude, direction and the sense in which they act i e $\overrightarrow{AB}$ or $\overleftarrow{AB}$. All can be represented by directed segments.

often referred to as representing a directed number (Figure 23). Now if directed numbers are going to be of use to us they

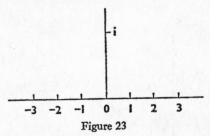

Figure 23

must conform to the rules of simple algebra (generalized arithmetic). For example, the following must be true:

1. The number $(3, 0)$ multiplied by $(4, 0)$ must be equal to the number $(12, 0)$. We must multiply together the lengths of directed number in the normal manner.
2. When the number 1 is multiplied by i the result must be i.
3. From No 2 it can be seen that multiplying the directed number $(1, 0)$ by i turns it through 90° and it becomes the number $\left(1, \dfrac{\pi}{2}\right)$, that is, it has unit length in the direction $\dfrac{\pi}{2}$. To be consistent, if we multiply the directed number $\left(1, \dfrac{\pi}{2}\right)$ by i it must turn through a further 90°, and this new directed number will be of unit length in the negative direction of the starting line; that is, it will be the number -1.

Hence $(1 \times i) \times i = -1$

Or $i^2 = -1$

We can also express the result of No 3 as:

$$\left(1, \frac{\pi}{2}\right) \times \left(1, \frac{\pi}{2}\right) = (1, \pi)$$

this being a particular instance of the generalization

$$(1, \theta) \times (1, \Phi) = (1, \theta + \Phi)$$

143

To multiply two directed numbers we multiply the lengths and add the angles. Multiplication of directed numbers is commutative.

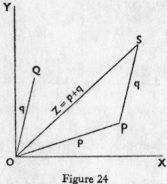

Figure 24

Directed numbers are added in the same manner as vector quantities, e g velocities or forces. Let p and q be two directed numbers represented by $\overrightarrow{OP}$ and $\overrightarrow{OQ}$. From P mark off $\overrightarrow{PS}$ equal and parallel to $\overrightarrow{OQ}$. Let $\overrightarrow{OS}$ represent the directed number z. Then the sum of p and q is *defined* as $z = p + q = q + p$. The addition of directed numbers is also commutative (Figure 24).

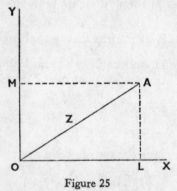

Figure 25

It is now possible to discuss the components of a directed number. Let $\overrightarrow{OA}$ represent the directed number z, and let

144

x, y be the co-ordinates of A. Further, let L and M be the projections of A on OX and OY. Then since $OL=x$ and $OM=y$ in magnitude and direction, one has in the direction of directed numbers:

$$x=\overrightarrow{OL} \text{ and } yi=\overrightarrow{OM}$$

But as has just been indicated $\overrightarrow{OA}=\overrightarrow{OL}+\overrightarrow{OM}$.

Therefore $z=x+yi$

Hence every directed number is the sum of two numbers, one number being directed along $\overrightarrow{OX}$ or $\overleftarrow{OX}$ and one number along $\overrightarrow{OY}$ or $\overleftarrow{OY}$. These two parts were originally called the real and imaginary parts, as was seen earlier. Although in the modern approach to mathematics these names are not really appropriate, they have remained. If neither x nor y is zero, then z is obviously a *complex* number (Figure 25).

Ordinary arithmetic and algebra, i e a logical and consistent set of rules for calculation, enabled us to represent real numbers as displacements on a line in terms of a single quantity. One has now made the algebra more general by making it possible to deal with displacements in a plane.

ALGEBRAIC AND TRANSCENDENTAL NUMBERS

Earlier real numbers were divided into two classes of numbers —the rational and the irrational. But one can also divide them in another way to give two other classifications that are significant—i e *algebraic* and *transcendental* numbers. The real number α is said to be algebraic if there is some equation

$$p_0x^n+p_1x^{n-1}+ \ldots p_{n-1}x+p_n=0 \ (n > 1)$$

with integers p_0, $p_1 \ldots p_n$ as coefficients, having α as root. Every rational number of the form $\frac{p}{q}$ must be algebraic, since it satisfies the equation $qx-p=0$. But any real number that is not algebraic is said to be transcendental.* From the above definition it is not easy to see that there are any transcendental

* Euler said that such numbers 'transcend the power of algebraic methods'.

numbers, although it has been known since about the middle of the 19th century that there are such numbers. For example, π is a transcendental number, and logarithms, to base 10, of most numbers are also transcendental.

It is possible to extend the above classification to complex numbers. One may define more generally an algebraic number as a number x, real or complex, that satisfies the equation

$$p_0 x^n + p_1 x^{n-1} + \ldots p_{n-1} x + p_n = 0 \ (n > 1)$$

if the coefficients are integers: and any complex number which is not algebraic is defined as a transcendental number.

REFERENCE

SCHAAF, W L (1960). *Basic Concepts of Elementary Mathematics.* London: Wiley.

CHAPTER TWELVE

Some Final Comments

It will be seen that the growth of the basic mathematical and scientific concepts that have been discussed is a slow and complex process, which is not, as yet, well understood. It does seem certain, however, that the concepts do not develop in an 'all or none' fashion, but rather that they are at first somewhat vague and hazy concepts, which grow in clarity, in width, and in depth, with maturation and experience. The rate of development seems to depend on the quality of the brain mechanisms of the child, upon his motivation, and upon the cultural milieu (which includes conditions within the classroom).

Before systems of actions carried out in the mind (i e systems of operations) can be stable, they must, according to Piaget, have certain properties; unless these are present, thought will show inconsistencies. The properties are (compare the properties of the mathematical group):

1 closure.—Any two operations can be combined to form a third operation. E g all boys and all girls equals all children; $3+4=7$.

2 reversibility.—For any operation there is an opposite operation which cancels it. E g all boys plus all girls equals all children, but all children except all boys equals all girls; $3+4=7$ and $7-4=3$.
   associativity.—When three operations are combined it does not matter which two are combined first; or the same goal can be reached by different routes. E g all adults plus all boys and girls equals all boys plus all girls and adults; $(1+4)+2=1+(4+2)$.

4 identity.—This is a 'null operation', and is performed when any operation is combined with its opposite. E g all human beings except all those who are humans beings equals nobody; $5-5=0$.

Now it will be remembered that addition of number is the fundamental process by definition, and subtraction the *inverse* of addition; just as multiplication (continued addition) is a fundamental process and division its *inverse*. When the child is nearing the stage where systems of operations in relation to the natural numbers are becoming stable for him he will *understand* that $3+4=7$ and $7-4=3$ are different ways of expressing the same relationship. Before then these may be discrete and unrelated facts for him.

The problem for the teacher is to ensure that the number bonds can be used in the mind as operations. Sometimes an attempt is made to bring this about by rote learning, although it is by no means certain that, because a child can use and/or verbalize the bonds, he can manipulate them operationally, using the term in a Piagetian sense (i e manipulate in the mind). Again, children may acquire partial or limited insights using, say, concrete material or real life situations, yet the insights may never link with other insights to form reversible operations. For example, here are three problems involving natural numbers:

'By how much is 37 greater than 19?'
'How many must be taken from 37 so that 19 are left?'
'How many must we add to 19 to make 37?'

Now, assuming the child can read, and he can comprehend what is asked of him, he should be able to grasp that subtraction is involved in all these problems; and so is addition in the sense that one number involved is the sum of the other two. If he does not understand these relationships it is unlikely that his thinking about number has reached the stage of reversible operations.

In the case of slow learning children, who need concrete materials up to 10 or 11 years of age or longer, it is important that teachers do not overestimate their capacity for number operations. They may well acquire concepts of sufficient width and depth for some real life arithmetic, or acquire rote knowledge and skill that can be made available for buying, selling, measuring, writing out simple bills and so forth, but they may never be able to use number in operational fashion.

It is vital that slow learning children be helped to understand whatever they can, but teachers must not delude themselves as to the limits of their pupils' understanding.

Eifermann and Etzion (1964) point out that adults find addition easier to perform than subtraction although they undoubtedly understand their reversible nature. In a small experiment they created a system isomorphic to the number system of from 1 to 10, based on the differences between various lengths of coloured rods. When the system was taught no preference was given to addition over subtraction, and it was found that there was no significant difference in the time taken by the subjects to carry out the operations of addition and subtraction. This confirms experimentally what good teachers have preached, namely, that emphasis should be placed on the reversible nature of the operations.

The capacity of the individual to form stable systems of operations in relation to number is obviously linked with his capacity for generalization and transfer of training. In the case of a child who sees no connection whatever between 16×5 and the problem 'If I share 80 oranges equally among five children how many will each get?', it is unlikely that his capacity for generalization is very great. In all mathematics one should teach for generalization, that is, one should constantly present to pupils, as far as is possible, new data as showing the validity of old principles.

Dienes (1959) has suggested that, if a concept involves a number of variables, to ensure understanding all the variables should be altered in turn. This is absolutely true, but a warning must be given to the effect that the junior school child has a limited capacity for varying one factor at a time, holding others constant, and appreciating what he is doing. Inhelder and Piaget (1958) have made this clear, and it has been confirmed by the writer (Lovell 1961), using average and bright English junior school pupils. We must therefore, when working with 7–11-year-old children, proceed with care while using apparatus involving a number of variables. This limitation in child thinking might also help to explain why the junior school child has such difficulty in eliminating the variables that could possibly affect conservation of quantity, weight and so on.

Again, Dienes uses, among other apparatus, a balance to help in the teaching of linear equations. A suitable piece of wood is balanced about its centre, and it has numbered hooks placed at equal intervals on either side of the point of balance. Rings can be placed on the hooks. The child is told that the pull on the balance is given by:

(The number of rings) TIMES (the number over the hook).

After various examples the child is led to the point where he has to find the value of A in the equation $12 = A \times 4$.

In many ways this is an admirable lead into linear equations, but one must bear in mind that at the junior school stage even the bright child will 'balance' by trial and error. This has been demonstrated by Inhelder and Piaget (*op cit*) and confirmed by Lovell (*op cit*). At this age and until about 12–13 years of age co-ordination between weight and distance is usually achieved by trial and error. The pupil may realize that there is a qualitative correspondence between weight and distance; i e 'the heavier the weight is the closer you put it to the centre', but he does not yet possess any concept of metrical proportion, because the *schema* of proportionality is not yet available to him. There is nothing whatever against the teacher using the piece of apparatus as long as he realizes the limitations of the junior school child's thinking, and grasps that the pupil's understanding may not be as great as it appears to be. Indeed, as Lunzer (1956) pointed out, it may be worth while for a teacher to go through processes where the child does not quite understand rather than wait until the solution is readily seen, because he is thereby increasing the child's familiarity with the material and the kind of reasoning involved. The value of familiarity with the type of problem is a point also made by Peel (1960, p 165 f, p 181 f) and Hindham (1960).

The aim of the teacher is to ensure that the concepts discussed do in fact develop. Furthermore, the speeding up of the process of development would also be an aim of many teachers. We do not know, however, how best to realize these aims. It may well be that there is no one best method, procedure, or piece of apparatus for all pupils. The two most important factors, *external to the child*, likely to affect the development of mathematical concepts are:

(*a*) the mathematical understanding of the teacher, and

(*b*) the climate of opinion in which the child is reared. This includes the way the teacher discusses and demonstrates such concepts in the classroom. (Naturally, the finding of the current N F E R research may show the superiority of one approach to the teaching of number and number operations).

In respect of (*b*) we have already said that the farther mathematical ideas obtrude into everyday life and experience, the more likely it is that children will imbibe them to an increasing extent. Goethe called the unconscious influence of current belief upon thinking the *Zeitgeist*—the spirit of the times that changes with the times. It is likely that a greater understanding of mathematical concepts is spreading throughout the mass of children to an extent unknown to earlier generations, and this is likely to be attributable in part to the culture pattern and not wholly due to compulsory schooling. One should not underestimate the effect of the *Zeitgeist*—the anticipations and synchronisms of thinking. It seems as if it was this that caused Napier and Briggs to invent logarithms in 1614; Newton invented the calculus in 1671, Leibnitz a little later, Newton's teacher Isaac Barrows having partly anticipated both.

A better understanding of mathematical concepts may be expected as the mathematical knowledge of our teachers improves, and as the children's experience becomes richer in relation to these concepts. One might expect a slow and steady improvement rather than a sudden growth due to this or that piece of apparatus, although all the materials discussed, and others, can in themselves provide learning situations. There is some danger that using specially constructed apparatus we may try to force a concept on a child before he is ready for it. If so, there is little hope of the concept becoming more generalized or remaining stable when the apparatus is withdrawn. Provided one does not, when using special apparatus, overestimate the level of the child's thinking and conceptual development, no harm will be done to ourselves or to him.

Earlier, stress was placed on the great importance of experimentation and experience in many situations and media for the development of the concepts of substance, weight, time, and space.   On the other hand, there is undoubtedly some lower limit, different for each individual, below which it is impossible for the child to begin to understand the ideas involved.   Lunzer (1960) has suggested why simple conservations, classifications, and seriations so often appear round about the same age, 7 to 8 years.   He suggests that the same psychological mechanisms are involved in all these concepts, namely, the ability to compare two judgments simultaneously. This is not the same as the ability to consider two perceptions simultaneously, for the latter amounts only to a simple judgment.   When the child can hold two judgments in thought, simultaneously, he is liberated from the here and now—that is, from his perceptions. But because his experience varies in respect of play situations and materials, this ability will not develop evenly in all fields, or even within the same field, at the same time.

A number of studies have been undertaken to see if intensive periods of specific training can speed up the growth of the understanding of a particular issue, e g conservation or transitivity in 5 to 7-year-old children.   The results have been faithfully noted in this book as given by the various authors of the published papers.   In the present writer's judgment, however, the overall of the effect of these training programmes has been small.   At most the training lasted only a few weeks.   Piaget's view is that the child may learn something from a specific situation, but it will have no effect on the child's general level of understanding for the specific attack is too trivial (compare Beilin 1965, 1966). The modification of a child's mental structure necessitates a far wider, more lasting and more radical approach which involves all the child's activities.   Phemister (1962) indicated that in free play that involved certain contrived situations, and which extended over five months, conservation of number might be helped forward.   The numbers of children involved were small, and a larger experiment is needed before too firm conclusions are drawn.   But Phemister's approach is more likely to produce lasting effects than is a training period of a few weeks.

It must be recognized, alas, that many children do not make much progress in understanding number and number operations and the other mathematical and the scientific concepts discussed. And it is often asked if there is any relationship between such general personality variables as instability and anxiety, and performance in number work. There are, of course, many children in the schools who are clinically within the bounds of normality and yet may show slight instability or anxiety. Do children who have these personality traits perform poorly in mathematics? Biggs (1959) has made a survey of the literature on this problem. There is a little evidence that certain general personality traits, such as instability and anxiety, to some extent prevent a person from realizing his full cognitive potentialities in many different fields of learning, including mathematics. Indeed, the relationship between maladjustment and poor performance may be closer in mathematics than in other subjects, although we should not be too dogmatic about these matters. If temperament is responsible to some extent for poor number ability in some children, failure may make the slight maladjustment worse. Again, during a broadcast talk, Biggs (1962) suggested that in associative or rote learning of number, motivation has to be supplied externally, e g praise or blame by the teacher. Moreover, all external motivation is based primarily on anxiety and this can be harmful if it rises above certain level. On the other hand, Biggs contends that using structured material the child abstracts the basic essentials of the concept for himself, and when this is the case the child needs no pushing, for he will want to know for himself how the concept may be applied and extended. As Biggs sees it, this difference in motivation is the key to the problem of emotional blockages in mathematics. In the research now being conducted by the National Foundation for Educational Research an attempt is being made to answer some of the questions relating to anxiety and performance.

It is necessary to know precisely how psychological stress affects perception and learning. For example, how do fear, anxiety, the effects of continual failure and the effects of parental discipline and attitudes, make their presence felt? How do we ignore signals that cause distress or excitement? How do expectancy, habit, and mental set exert their influence? Is there some selection or filtering of signals in the centrencephalic system—that central system within the mid brain which has been, or may be in the future, demonstrated as responsible for the integration of the functions of the hemispheres, or the integration of varied specific functions from different parts on one hemisphere—or does the filtering or selection take place between the relevant receptor organ and the centrencephalic system? These are some of the questions that need to be answered.

Skemp (1961) has produced evidence suggesting that the transition from elementary number work to mathematics involves the use of *reflection* on the relevant concepts and operations. If an individual does badly at mathematics it may not be because he has not formed the necessary concepts and operations, but because he cannot reflect on them, i e perceive the relationships between the concepts and operations and act on them in ways which take into account these relationships and other relevant information. Skemp hints that this reflective ability may be affected by emotional disturbance. Clearly more research is needed here. We must be very careful that we do not label as emotionally disturbed those who find mathematics difficult; outstanding mathematicians have been among those who have had periods of great disturbance in the emotional life.

CONCLUDING COMMENTS

The great work of Piaget has shown how logical thought, i e reversible thought forms that give thought its internal consistency, grows out of the interaction of the organism and the matter-energy system (physical universe), aided by certain kinds of linguistic structure and cultural milieu. The motor action leads to mental operations, and the grouping and co-ordinating of these manifests itself in what we call *intelligence*.

Logical thought is the most powerful intellectual tool man has for coming to grips with the matter-energy system. Unfortunately men possess the capacity for such thought in very varying degrees—many seem to have but little of the capacity. It is not until mental operations are developed and co-ordinated, as the result of action and experience, that the individual can understand his environment. We cannot, as it were, ever 'teach' children number, length or time in isolation. Finally, while logical or internally consistent thought is so important to man we ought not to overestimate its applicability. Experi-

FINAL COMMENTS

mental psychology has again raised in acute form the problem: What relevance, if any, has logical thought to the manipulation of concepts dealing with data considered to be without the matter-energy system?

REFERENCES

BEILIN, H (1965). 'Learning and Operational Convergence in Logical Thought Development'. *J Exp Child Psychol*, **2**, 317–339.

BEILIN, H (1966). 'Feedback and Infralogical Strategies in Invariant Area Conceptualization'. *J Exp Child Psychol*, **3**, 267–278.

BIGGS, J B (1959). 'The Teaching of Mathematics—II: Attitudes to Arithmetic—Number Anxiety'. *Educ Research*, **1**, No. 3, 6–21.

BIGGS, J B (1962). 'The Psychopathology of Arithmetic'. *The Listener*, 19th April, 681–683.

BIGGS, J B (1962). *Anxiety, Motivation, and Primary School Mathematics*. NFER Occasional Publication No 7.

DIENES, Z P (1959). 'The Growth of Mathematical Concepts in Children through Experience'. *Educ Research*, **2**, No. 1, 9–28.

EIFERMANN, R R and ETZION, D (1964). 'Awareness of Reversibility: Its Affect on Performance of Converse Arithmetic Operations'. *Brit J Educ Psychol*, **34**, 151–157.

HINDHAM, Y H (1960). 'The Effect of Familiar and Unfamiliar Settings on Problem Solving in Mathematics'. Unpublished Ph D Thesis: University of Birmingham Library.

INHELDER, B and PIAGET, J (1958). *The Growth of Logical Thinking*. London: Routledge and Kegan Paul.

LOVELL, K (1961). 'A Follow-up Study of Inhelder and Piaget's *The Growth of Logical Thinking*'. *Brit J Psychol*, **52**, 143–154.

LUNZER, E A (1956). 'A Pilot Study for a Quantitative Investigation of Jean Piaget's Original Work on Concept Formation: a Footnote'. *Educ Review*, **8**, 193–200.

LUNZER, E A (1960). 'Some Points of Piagetian Theory in the Light of Experimental Criticism'. *Journal of Child Psychology and Psychiatry*, **1**, 191–202.

PEEL, E A (1960). *The Pupil's Thinking*. London: Oldbourne Press.

PHEMISTER, A (1962). 'Providing for "Number Readiness" in the Reception Class'. *National Frobel Foundation Bulletin*, April, 1–10.

SKEMP, R R (1961). 'Reflective Intelligence and Mathematics'. *Brit J Educ Psychol*, **31**, 45–55.

INDEX